In No Time

The
Internet

IN NO TIME

The
Internet

Ingo Lackerbauer

Edited by
ROB YOUNG

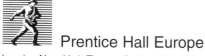

Prentice Hall Europe

London New York Toronto Sydney Tokyo Singapore Madrid Mexico City Munich Paris

First published in 1997 as Easy – Internet by
Markt&Technik Buch- und Software Verlag GmbH
85540 Haar bei München/Germany
This edition published 1999 by
Prentice Hall Europe
Campus 400, Maylands Avenue
Hemel Hempstead
Hertfordshire, HP2 7EZ

A division of
Simon & Schuster International Group

© Prentice Hall Europe 1999

Translated by Mark Finniear and Anne Hulme
in association with First Edition Translations Limited, Cambridge

Typeset in Stone Sans
by Malcolm Smythe

Designed and Produced by Bender Richardson White

Printed and bound in Great Britain
by TJ International Ltd., Padstow, Cornwall

Library of Congress Cataloging-in-Publication Data
Available from the publisher
British Library Cataloguing in Publication Data
A catalogue record for this book is available from the British Library
ISBN 0-13-977661-3

1   2   3   4   5   02   01   00   99   98

Contents

2 Making your PC fit for the Internet — 46

3 Onto the World Wide Web — 76

Loading a file from the Internet 198

10 Discussions on the Internet: Newsgroups 214

Dear readers,

In recent months and years, an increasing number of people have discovered a new hobby – virtual travel on the Internet. Online surfers bridge thousands of kilometres on their PC within seconds, without even setting foot outside the door. These journeys lead to the discovery of new areas of knowledge and new friendships. There is an element of discovery if you probe the depths of the Internet, time races by and, hours later, reality suddenly dawns on you again.

This book is directed towards those people who would like to learn how to handle the modern communications possibilities in an entertaining way and discover a new world for themselves. Beginners will be taken by the hand and gradually initiated into the new possibilities and ways of the Internet. The book is intended to give them the most important basic elements for using and operating the Internet and the most important Internet services along the way. Precedence is given to the often quoted motto 'learning through practice', since only in this way is it possible to enter the online world enjoyably and through experimentation.

I hope you enjoy working through this book and surfing online on the Internet and that you will learn a lot and many new friends along the way. If you would like to contact me, simply send me an e-mail on ingo@ito.de.

See you online!

Ingo Lackerbauer

The following three pages show you how the computer keyboard is constructed. Groups of keys are dealt with individually to make it easier to understand. Most of the computer keys operate in the same way as on a typewriter. There are also some additional keys, however, which are designed for the peculiarities of computer work.

See for yourself . . .

Typewriter keys

You use these keys exactly as you do on a typewriter.
You also use the Enter key to send commands to the computer.

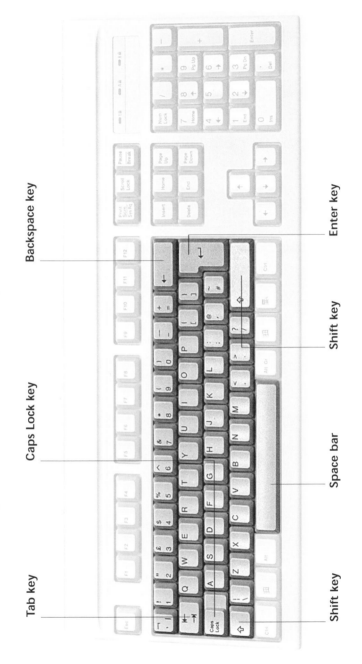

Backspace key

Enter key

Caps Lock key

Shift key

Tab key

Space bar

Shift key

13

Special keys, function keys, numeric keypad, status lights

Special keys and function keys are used for particular tasks in computer operation. The `Ctrl`, `Alt` and `Alt Gr` keys are mostly used in combination with other keys. The `Esc` key can be used to cancel commands, and Insert and Delete used, amongst other things, to insert and delete text.

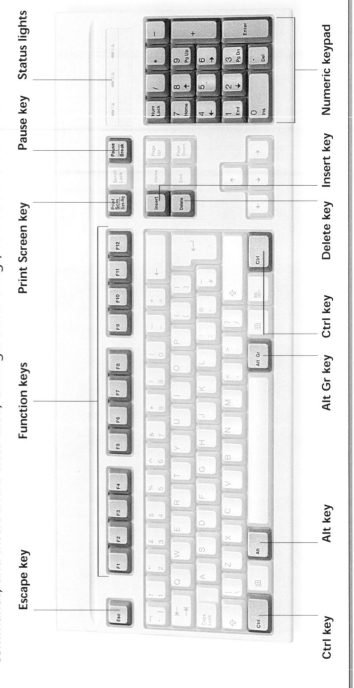

| Escape key | Function keys | Print Screen key | Pause key | Status lights |

| Ctrl key | Alt key | Alt Gr key | Ctrl key | Delete key | Insert key | Numeric keypad |

Navigational keys

These keys are used to move around the screen.

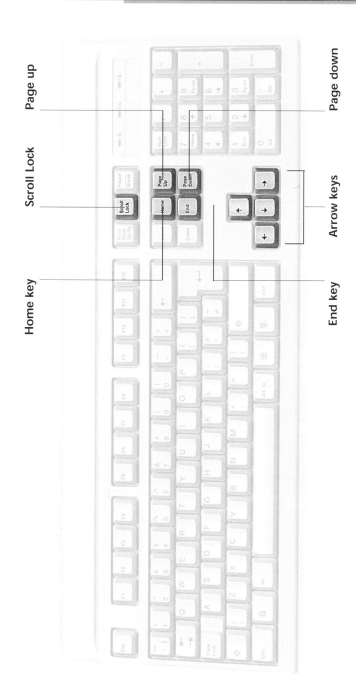

Page up

Scroll Lock

Home key

Page down

Arrow keys

End key

15

Clicking with
the left-hand
mouse button

'Click on . . . '
means: press once
briefly on the
button.

Clicking with the
right-hand mouse
button

'Double click on . . .'
means: press the left-
hand button twice
in quick succession
very fast.

Double clicking

'Drag . . .'
means: click on an object with the left-
hand mouse button, keep the button
pressed, move the mouse and
thus drag the item to
another position.

Drag

What is the Internet?

What's in
this chapter?

In this chapter, you will learn everything
about the history of the Internet and its
structure. You will also gain all the necessary
knowledge to enable you to log onto the
Internet as well as discovering what you can
do on the 'network of networks'. Finally, you
will learn about the differences between the
Internet and an
online service.

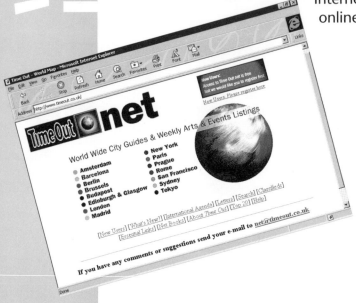

Your are going to learn:

History and structure of the Internet

The Internet has become as common as the daily newspaper, television, radio or video. It is emblazoned on advertising hoardings, advertised on TV and has huge press coverage devoted to it. It is reputed to follow the steam engine and electricity in changing all our lives fundamentally. Hardly a TV programme goes by, without the presenter referring to an Internet address, where you can find out more about the show or respond in some way as a viewer. You can lie comfortably on your sofa at home and travel on the Internet to the furthest corners of the earth, without even setting foot outside your front door. You can plan **journeys** and make reservations over the Internet, find out the latest **football results**, consult **cinema listings**, explore **books**, listen to new **CDs**, **chat** with like-minded people in any part of the world or call up inexhaustible **data sources** on a chosen topic. There is nothing you cannot find on the Internet! The possibilities which this new medium offers are as wide-ranging as the people themselves with all their interests and tastes.

The Internet is a microcosm of the real world around us. To define the Internet is as difficult as to describe the world in which we live. It can best be compared to a **market place**: anyone can come into the Internet market place, be they spectators, sellers or customers – regardless of whether they are Joe Bloggs or the Prince of Wales. There are no class distinctions in this medium. Everyone has the option just to look at something, buy something or perhaps even offer their own wares for sale.

The Internet will change the world and is already doing so. The future significance of the Internet will surpass anything to date; it is simply becoming the 'global communication medium'. These grand words perhaps sound a little dramatic they are but accurate. The sweeping success of the Internet will not only change **professional life** in the coming years but will also shape the future of commercial products, business enterprises and major companies. Experts say we are on the brink of a **revolution**, that will change our lives even more than the industrial revolution did 200 years ago. We must simply wait and see what happens! The butcher's corner shop will be less affected by it, for example, than large conglomerates spread right across the globe. The technology has penetrated even the butcher's life, however, which would have been absolutely inconceivable just 15 years ago. Today, you'd hardly find a butcher without a PC! Computer technology took 10 to 15 years to gain a foothold in society – the Internet will certainly not take so long. By the new millenium, the Internet will be as widespread and accepted as the **telephone** and television. There are already efforts to integrate the Internet into telephones and **televisions**, so that PCs will no longer be necessary to surf the Internet.

At the start of the nineties, only professors, students and the initiated knew the term 'Internet'. Just eight years later, the Internet has become a central cultural and economic topic. A small press article from November 1996 shows how far the Internet has already become part of everyday life in the USA:

The presidential elections in the USA have broken all records on the Internet: Millions of users followed the count online. The CNN web site alone received 50 million visitors in 24 hours. America Online, the largest online provider in the USA, meanwhile registered over 155,000 requests per minute. (DPA)

You can see therefore, how the **information superhighway** will affect your life – whether you like it or not! You will chat to friends and acquaintances on the Internet, send letters, go shopping, catch up with the news, solve problems, play, watch TV and – if you want – earn money!

The Internet is the ideal medium for the new millenium – fast, cheap and effective.

21

What is the Internet – technically speaking?

The Internet, from a technical perspective, is an interconnection of millions of computers throughout the world. Computers in over **150 countries** around the world are currently connected to the Internet. The Internet connects mainframes and PCs, which are held by governments, the military, academic institutions and private individuals.

History of the Internet

Anyone who thinks that the Internet is a 'discovery' of the nineties, is mistaken – and seriously so! It all started in 1957, when the USSR sent the first **Sputnik** satellites into space. With this technical masterstroke, America's worst fears became a reality: their adversary in the Cold War had gained supremacy over space! Alarm bells sounded loud in the **Pentagon**. The RAND Corporation, America's premier think-tank during the Cold War, was faced with a particular strategic problem. How could the American government communicate successfully after a nuclear attack? A computer centre with a massive collection of all 'high tech' computers of the day could be armed and protected but the lines could be destroyed by a **nuclear attack**.

A **network**, also termed a LAN (Local Area Network), connects several computers to one another via a cable. The combination of several networks, which are spread right across the globe or located in various cities, is termed a WAN (Wide Area Network).

The staff at **RAND** pondered and brooded, until they finally arrived at a daring conclusion: firstly, a network should no longer be centrally controlled. In practice, this meant that there should no longer be a main computer centre in which all the threads come together. The new type of network was also intended to remain operational in **fragments** under RAND's proposals. The principle is simple: the individual computers on the network, which would be spread right across the USA and all connected to one another, all have the same status in relation to one another in sending, forwarding and receiving messages.

A **packet**, also termed data packet, is the information, that is waiting to be sent across a network. The pieces of information, such as electronic mail, for example, are 'tied together' into small manageable units, in order to be able to transmit these easily across the network cable.

The messages themselves are broken down into small **packets**. An address is attached to these small data packets (as in a post office). All packets start from a common computer and are combined at the destination once more into a single piece of information. Each packet finds its own way through the network, with the only objective being to reach the addressee. The path of each individual packet is unimportant, only the result matters. If a large part of the network were destroyed, this would not matter, since the message always finds a path to the destination. This also results from the fact that each computer or each network is connected to all others – similar to a **spider's web**.

In the sixties, this **bombproof** packet-sending network was proposed by the RAND Corporation to various national institutions, which were naturally more than enthusiastic about the abilities of this network. Just a short time later, the USA decided to establish the 'Advanced Research Project Agency' (**ARPA**), which was to be directly answerable to the American Department of Defense (DoD). ARPA thereupon initiated an even bigger and more ambitious project and then slowly switched from the military network to scientific use. In autumn 1969, the first supercomputer (or what was then considered to be such) was installed at the University of Los Angeles, with the aim of constructing a network, that was to operate according to the specifications of the RAND Corporation.

In December 1969, there were already a total of four nodes on a small network, which spread across the USA and was named after its Pentagon sponsor. Thus, the **ARPANET** was born.

A computer in a network is also termed a **node**.

Thanks to ARPANET, scientists and researchers could use the computer facilities of other institutions even over long distances. This was very useful, since a lively exchange of information could then take place between the universities. From then on, it went from strength to strength: in 1971 there were already 15 nodes on the ARPANET, and by 1972 there were 37. In the seventies, the ARPANET continued to grow. By 1983, the ARPANET structure had grown into an international network, to which an increasing number of universities and colleges were connected. In 1983, it became a little too colourful for the sponsors from the Pentagon and ran the risk of supervision being lost. This was the reason for breaking up the ARPANET into the MILNET, which served military communication, and the original ARPANET for further research facilities. Researchers and engineers worldwide developed networks, such as CSNET or BITNET, which were also intended to connect to the ARPANET. Private companies also saw the advantages of networking and sought access to the ARPANET. Although the ARPANET was growing, it was disappearing increasingly in the plethora of worldwide networks. Thanks to a uniform standard in packet transmission, ever more networks throughout the world joined the ARPANET. The Internet was born. Today, decades later, there are more than **16 million** computers connected to it in over 150 countries. Insiders predict that by the year 2000, more than **250 million** subscribers will be using the Internet.

In 1984, the 'National Science Foundation' (NSF) took a hand in pioneering a step in the direction of technological progress. The NSFNET arose out of a particular requirement. The NSF wanted to connect its six supercomputer centres in the USA and developed its own particular network programs for this purpose. From 1986 onwards, the NSFNET was continuously extended, since the data traffic on the NSFNET increased considerably (10% per month). Meanwhile, the NSFNET had become a national research network. Over the course of time, national institutions such as NASA, the National Institute of Health and the Department of Energy also connected up to this network. The ARPANET was dissolved in 1990 having almost become a victim of its own success. Its structures were integrated into the larger Internet.

In 1991, the NSFNET was renamed NREN (National Research and Education Network). But the regulation from the NSFNET period, that no commercial data could be transmitted over the NSFNET, remained. This regulation expired in 1995, however, so that it is now possible to transmit commercial data in addition to research and education data.

The Information Highway

In 1991, the NREN was agreed by the American **Congress** as the official successor to the NSFNET. Since the child had to have a name, they settled on 'High Performance Computing Act' (HPCA). HPCA is a US government promotion programme which subsidises the high-speed networks with more than US $1 billion. The results which arose in the framework of this research are intended to feed into a new, National Information Infrastructure (NII). The American vice-president, Al Gore, declared the NII initiative to be at the top of his agenda. Politicians see the aim as enabling access to multimedia communication (videophones, **TV shopping** etc.) for everyone in the USA. The Internet acquires a central role at this point. It has come of age and will assume a central role in our lives in the coming millennium. We wait in eager anticipation!

Who runs the Internet?

Beginners and newcomers to Internet issues pose this question over and over again. Is it powerful firms or company bosses who run the Internet, demanding large sums of money from data travellers and controlling the Internet structure? Not a bit of it. The network of networks, as you have already discovered, is constructed in a decentralised way and has anything but a rigid structure. Rather, we are dealing here with a large, globe-encompassing web, which unites hundreds of thousands of computers and innumerable subscribers under one roof. On the Internet, everyone is his or her own boss and/or operator. As you can imagine, such a structure is chaotic. There is noone at the head of the Internet who can issue general instructions or can pass Internet laws.

If the operators of an Internet computer decide to detach their computer from the Internet, this is far from being a disaster, since the data will immediately find a different path to the destination thanks to the woven meshed network.

The Internet is not based on the **involvement** of professional businesses, but instead on the many thousands of information providers. The main task of the Internet falls on the shoulders of the many unknown individuals located at universities, in offices or in their own homes who provide the Internet with information – and do so free of charge. Many private Internet enthusiasts help in their free time to add new servers to the Internet, improve existing servers or correct any faults.

As soon as you are on the Internet, you will therefore not be using the service of a large company or even of a state **organisation** but various services of different people and institutions who all have the same rights and obligations on the Internet. The lion's share of the **costs** and work required for the maintenance of the Internet is borne through the time and financial resources contributed by them, with the result that this is not included in the actual cost calculation for your Internet access. Thanks to these decentralised and partly anarchic structures, costs are incurred only for access to the Internet and **telecom charges** – but more on that later.

The basics of the Internet

Before you delve deeper into the Internet, you should learn some terms that you will encounter time and again on your excursions.

WHAT'S THIS?

Protocol: A protocol describes a set of rules according to which data is transmitted over a medium, for example the Internet. This might be compared with the Highway Code. A protocol therefore ensures that all data is correctly sent and combined again at the recipient.

The Internet is inseparably associated with the terms **TCP/IP** and **IP**: This is the Internet transmission protocol, which regulates the exchange of data on the Internet from computer to computer.

IP address: Every computer on the Internet and other devices, such as Internet cameras and printers, for example, have an unambiguous number, by which a computer can be identified.

For smooth functioning of such a complex structure as the Internet, clear guidelines are required for the identification and naming of devices (computers, printers, Internet cameras, etc.), which are found in the Internet structure. Every resource on the network therefore has an unambiguous number – the **IP address**. This number has the form aaa.bbb.ccc.ddd and always consists of four numbers. Each individual number may have a maximum of three digits and values between 0 and 255. These regulations were developed by the founders of the Internet and have been retained down the years. A possible IP address might be 194.167.25.23, for example.

Home page: Any private person or company represented on the Internet or WWW with its own pages has one page that first welcomes the visiting user – the home page.

Since these numbers are difficult to remember, FTP and WWW servers (more on these later) have a more meaningful name. Microsoft's **home page** on the World Wide Web can be reached via *http://www.microsoft.com*, which is considerably easier to remember than one of the numbers described above.

A **DNS server** is a computer on the Internet, that contains the addresses of all computers, that can be reached on the Internet. DNS servers are spread right across the globe and update with one another at set intervals of time.

How does the Internet know which IP address corresponds to which name? The 'Domain Name System' is responsible for this name allocation, better known by its abbreviation **DNS**. Corresponding DNS servers, which are located on the Internet, regularly talk to one another, so that a new allocation is notified around the world in a few hours.

27

The individual components of a name are separated from one another by points. The left-hand components of the name describe, for example, the computer or company concerned. The further you move through the name to the right, the higher you find yourself in the organisational structure of the Internet. The component of the name on the far right defines the domains and describes the organisation or company category to which the operator of the web server belongs.

The following domains are currently in use:

Organisational domains	Meaning
com	commercial
edu	educational
gov	government
mil	military
net	network (Internet Service Provider)
org	organisation (non-profit-making organisation)

Geographical domains	Meaning
au	Austria
ca	Canada
ch	Switzerland
de	Germany
fr	France
uk	United Kingdom

There are many more geographical domains – one for every country actively participating in the Internet.

Gateway to the Internet – the Internet service provider

As already explained earlier, the Internet is a structure which is not subject to any authority or company but is controlled in a certain way by particular companies – the **Internet service providers**. This long and technical-sounding name simply means that a company, that smoothes your way onto the Internet. Such companies currently offer optimal access to the Internet and provide an extensive range of other services. Most companies (Demon Internet, Virgin Net and many others) have attractive offers intended to provide you with cost-effective surfing. Collect offers from the various providers to find out what **Internet access** is best suited to your requirements. Check out whether your preferred provider has sufficient **Internet access points** near you. You or your computer should call the Internet computer of the Internet service provider via the telephone line, to make a connection between your PC at home and the system of the provider. Since the computers of the ISP (Internet Service Provider) are linked directly to the Internet, you and your computer become elements of the Internet throughout the connection network. You can therefore send data onto or receive it from the network of networks unimpeded. The providers must naturally be paid for this service. Several years ago, **charges** for connection to the Internet via ISPs were relatively expensive and unattractive for private individuals. These costs did not stand in the way of the Internet hype, however. Today, anyone can rummage around on the Internet for a few pounds a month.

Another tip: To keep the running costs down, investigate the public utility organisations, which offer Internet access almost at cost price.

On top of the ISP costs there are only the telephone charges, when you connect your computer to the Internet Service Provider. For this reason, you should note that your ISP offers access connections which you can get at the local call rate. You thus keep the costs low and are not surprised by horrendous **telephone bills** at the end of the month.

How to keep your costs down

Always keep an eye on your costs! As regards your access charges to the Internet service provider and telephone expenses, it is up to you to make the best of it. Seek out access which is tailored to your personal **requirements** and represents the best available for your purposes on the Internet. Checkout in advance whether you intend to use the Internet only privately (most cost-effective) or professionally (significantly more expensive). Maybe you do not want to use all the services the Internet offers or conversely, would like the full range including all services. Try to schedule the trip onto the Internet in the **evening**, to keep the telephone costs as low as possible. You will develop a feel for the accumulating costs early during your time on the Internet. It is best to install one of the many **charges programs** (which you can also download via the Internet). These useful programs show you precisely the level of telephone charges you have accumulated and inform you if you exceed a particular limit, which you can set in advance.

What can you actually do on the Internet?

It should have become clear now how the Internet arose and how you can access it. The next question is sure to be, *Once I am on the Internet, what can I actually do?* Here is a general look at the individual Internet services you can use, and what they are for.

Electronic mail – email and mailing lists

People enjoy contact. Writing letters is one of the oldest forms of human communication and is extremely popular, even today. Letter writing gains a whole new dimension on the Internet. The electronic mail, which you have just written, reaches the recipient after just a few minutes – regardless of whether the recipient is in Australia or Alaska. The recipient can likewise reply without delay (provided that time differences allow), meaning that a lively exchange of information arises via **email**. The advantages of electronic mail are

obvious: rain, hail, snow, postal strikes or flat tyres on the delivery van are not problems encountered by email. The electronic letter selects a path along the various cables around the world and, despite all conceivable obstacles, finds its way to the recipient – unless the power is cut off. As for costs per electronic letter, email puts conventional letter traffic in the shade. An email normally costs precisely one **charge unit**, no matter to which corner of the world the email is sent. That is not all. With a few tricks up your sleeve, you have the possibility of sending multimedia messages with your email such as pictures, voice, sound, documents or video sequences – something that is not offered by any other current state of the art communication medium.

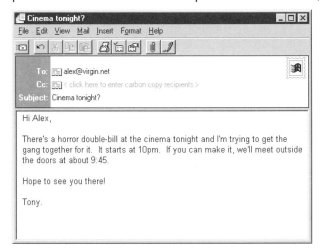

A particular form of Internet mail is the **mailing list**. This is a service which sends an email to all subscribers on a particular distribution list.

The **BITNET** is a network within the Internet, which is used primarily by colleges and universities around the world.

Mailing lists use the **BITNET** as their transport medium. This Internet subnetwork was started in 1981 to improve communication between universities and similar institutions. The BITNET has increasingly developed over the years, so that today, over 1,200 research stations around the world communicate via this network. As a result of this, many technical and scientific discussions are held at a very high level on the BITNET. Besides this scientific focus, there is also a great deal of information, that is of general interest.

The principle of the mailing lists is based on the collection and distribution of emails on various topics. The subscribers to a particular discussion group are entered in a list. To participate in this lively exchange of information, you only need email software capable of receiving Internet emails.

All news and messages that arrive dealing with a particular topic on this list are automatically distributed to all members of this mailing list and sent by email. If you need to write comments or a response to one of these articles, your email reaches the corresponding mailing computer, which then in turn forwards this to the relevant members of the mailing list. World-wide communication can thus arise, which eventually takes the form of a continuous discussion.

Mailing lists are not only represented, however, on the BITNET, they are also found outside the BITNET. But this is of minimal importance, since the Internet has connections with all networks using mailing lists. As a result, it is difficult to find out from what branch of the Internet your message originates. There is a big difference, however, in the management of the members and distribution of messages. The Internet mailing lists are not automatically processed, in contrast to those of the BITNET, but are the responsibility of an administrator, who observes that everything is in order.

Tips on writing an article

As in ordinary life, the Internet user must keep to certain **rules** and conventions in mailing-list matters. Many of the mailing lists have their own rules, which we will not go into here. You will very quickly learn the rules that apply to your mailing list. There are some general rules, however, which you should take note of at this point. After you have been accepted onto a mailing list, you should just read **articles**, in order to gain a feel for the medium. Avoid inundating the rest of the Internet community immediately with your knowledge – this might result in embarrassing situations in certain circumstances (bulls in china shops come to mind). In this way, you will avoid some of the mistakes that are frowned upon, and are often (but not always) made by beginners. One of the commonest of these is asking questions that have just been answered.

Mailing lists are like your local pub as far as the tone of the discussion and acceptable behaviour are concerned. Some discussion groups get excited, and speak bluntly, others are more formal or restrained.

A further discussion possibility within mailing lists is the **moderated discussion group**, which is based on strict rules and guidelines. In the framework of moderated discussion, your articles are not passed directly to members of the list but first read by a moderator and, if not suitable, discarded. What at first sounds very much like a censor, is intended to help to keep the number of circulating messages as small as possible by combining articles with the same content. The moderator can also keep the discussion on the topic and avoid digressions, in a similar way to the leader of a conventional discussion.

Topics to infinity – the newsgroups

A **newsgroup** describes a meeting place on the Internet where you can exchange views with other people on a particular topic.

The **newsgroups** are the most frequently used service on the Internet, besides the World Wide Web. So what are newsgroups? Imagine a magazine, that publishes all its readers' letters and makes these available to the general public, so that all readers can participate in the discussion. Now suppose that all magazines on every conceivable topic operated in this way and make their readers' reactions public. You probably think that such a magazine would be very interesting but hardly feasible in practice. On the Internet, things appear very differently. The newsgroups on the Internet offer this medium on an electronic basis. Newsgroups are groups (**discussion forums**) in which news is exchanged and discussions are held on a particular topic. Every Internet subscriber has the possibility of actively participating in the event by publishing text that can be read as electronic news by the rest of the Internet community.

The electronic messages of a newsgroup are also called 'articles' – and that makes sense. You browse through the news, in the same way as you would through a newspaper. When you find an area that awakens your interest, you can stop and read the relevant article. Most **articles** have a meaningful heading, so that you should know immediately what they are about. If you have an answer to a problem or a question, you can respond to this.

There are currently over 15,000 newsgroups on the Internet covering the widest range of topics – from A for automobile to Z for zeppelin. You find forums in the newsgroups, for example, on topics such as finance tips, legal information, travel, books, films, music, science and technology, the environment, politics, religion, computer topics and much more. So that you do not lose an overview of this range of topics, you need a program known as a **newsgroup reader**, with which you can quickly search for newsgroups, respond to questions and pose questions yourself.

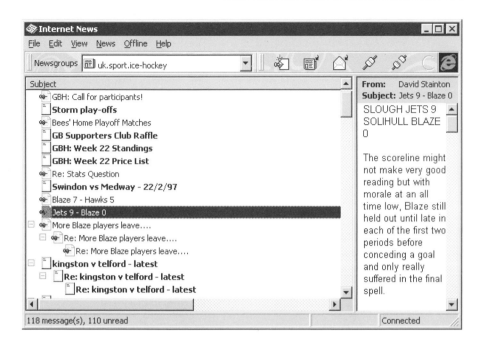

Since there are newsgroups on every conceivable topic, you will also find some on controversial topics such as sex. Recently, sex on the Internet has become a favourite subject for the tabloid press. This whole topic has been blown up out of all proportion. Certainly, there are forums, that deal with sexuality but these are no different from the other areas – there are harmless conversations and also serious discussions. Certainly there are also **black sheep**, and these are very closely monitored by the Internet service providers.

The exchange of information takes place within a newsgroup, by each subscriber sending a message as an email into the relevant group. All subscribers to the newsgroup concerned can read this mail. Your message also moves over the network from computer to computer, whilst the messages written there are in turn forwarded to the other computers. This has the consequence, that every computer connected to the Internet receives a copy of all discussion contributions written everywhere. The subscribers can read these or write new messages, and the circulation of the exchange of news

35

starts over again.

Paradise for the 'software hoover' – FTP

The Internet is paradise for people who like looking for new software. The Internet has software stretching as far as the eye can see. Whether it be programs for cooking recipes, music archiving or games, or clever programs designed to simplify daily life for you with your computer – you will find all that and much more on the Internet.

Viruses are programs, that can 'sneak' onto your computer and cause disastrous damage. There are harmless viruses, but also some that can format hard disks and engage in other unpleasant activities.

But beware! Before you start to download a program from the Internet onto your computer, you should definitely acquire a virus scanner. If you bring home **computer viruses**, you run the risk of endangering the health of your computer. These evil invisible programs, known as viruses, have the simple aim of putting your computer out of operation. Important data on the hard disk can be destroyed in this way or even the whole disk formatted. A **virus scanner** is a program that examines every program and every file for virus contamination. There are numerous such programs on the market, which are effective to varying degrees. Most of these useful aids are free of charge, in particular on the many CDs which are attached to the covers of specialist magazines. Once you have installed a program of this type and it is working correctly, there is nothing to stop you from **downloading** (loading a file from the Internet onto your computer).

What does FTP mean? There are a whole series of computers on the Internet that were specially established as software stores for the community of Internet users. Universities and manufacturers of computer components, in particular, offer one or more such systems, and almost all can be used free of charge by Internet users, to obtain the latest software down the telephone line. Many private individuals also run these types of computers to make their software available to

Freeware describes software which does not cost a penny and can be copied from the Internet free of charge.

Shareware, in contrast to freeware, is software that costs a small nominal fee, which must be paid after a particular period of testing the software.

the general public. This is primarily **freeware** or **shareware**, which can be transferred freely without infringing any licence conditions. Such a system is known amongst the Internet fraternity as an **FTP server**. A computer of this type serves you with files and programs. The abbreviation FTP stands for 'File Transfer Protocol', which actually means no more than the technical regulation under which a program should be deposited onto your computer from the Internet.

Naturally, in turn you need a program that allows you to download a file from the Internet using FTP. Windows 95 already contains such an application, which is not the most convenient form of FTP but totally adequate to start with.

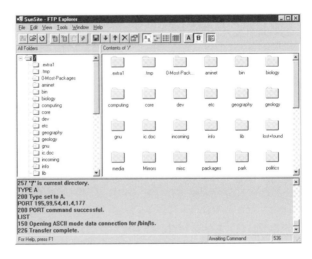

Working on the CIA computer – Telnet

Now it get's interactive! Telnet is a classic amongst Internet services, but is increasingly losing importance. In the early days of the Internet, there were no graphic user interfaces, such as the current Windows 95. To provide information for reading, **Telnet** was developed. This service allows you to log onto a remote computer via the Internet and work on this from the home environment. Due to new technology, such as the World Wide Web (more on that later), there are only a few Telnet Internet sites remaining, since many operators of Telnet servers have moved onto the World Wide Web. I have mentioned it at this point, for your information, in case the conversation ever turns to Telnet, even if it is no longer of any importance for you.

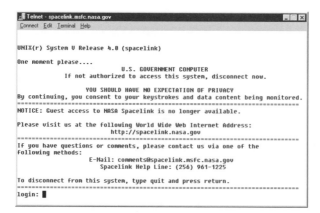

To select a Telnet server, you need a specific Telnet program again. As with FTP, Windows 95 comes with its own software which allows you to connect to Telnet computers.

Through the Internet at the click of a mouse – World Wide Web

As you are probably already aware, the Internet is no longer in the first flurries of youth. Most Internet services were developed at a time when there were no graphic user interfaces, such as Windows 95. This was also the reason why the Internet, until a few years, ago had an initiated community made up of only scientists, engineers and eccentrics. This changed rapidly in 1992, when the World Wide Web, also termed **WWW** or simply the **Web,** was established.

The advantage of this new service lies in its graphic user interface, which is simple enough for a child to use. The information accessible via the WWW has a **graphics outfit** and can easily be obtained using a mouse. Each computer on the Internet, which can be reached via the WWW, is described commonly as a 'WWW server' or a 'web server'. Each of these web servers has numerous pages, on which the surfer can find numerous pieces of information on every conceivable topic. These are generally produced by the operator of the web server and are looked after with greater or lesser degrees of care and attention.

To call up pages of a WWW server, you need corresponding software in the form of a **Web Browser,** or WWW browser. Well-known Web browsers include Microsoft Internet Explorer and Netscape Navigator.

It would be unfair to the World Wide Web to claim that this is only a new interface. The WWW represents a new Internet service with a wholly new feature offered by no other Internet service. In the past, there was no possibility on the Internet to link pieces of information with one another. This is precisely the possibility offered by the WWW. Instead of only showing text and graphics on one page, references can also be included at another point on that page, on another page or even on another WWW or FTP server. The WWW therefore gives you a free hand by simply clicking on a reference – also termed **hyperlink** or sometimes just link – to another page on the Internet, without having to enter a number abbreviation, other abbreviation or cryptic number/letter combinations. The WWW also includes other Internet services: a hyperlink on a web page can correspond to a file, for example, which can be downloaded via FTP.

Finally, another term that you should know – hypertext. Whilst normal text files are only linear and therefore processed in sequence, a **hypertext document** contains hyperlinks to other documents. Based on these, you are able to jump from one page to the next in any sequence and back again, in order to investigate another hyperlink. Hypertext contains not only text but also graphics, sound files and videos.

He who seeks, shall find – search services on the Internet

The Internet is immense and it is impossible to oversee the full range of information. As a consequence of this spread of information, it is becoming increasingly more difficult to filter out desired information from the heap of data. For precisely this reason, there are numerous **search services**, which are intended to help you in the search for information.

Archie

Archie is a classic help resource in searching for information. The volume of data that is available worldwide on FTP servers, is simply gigantic. The only problem is to track down available files on a desired topic in an

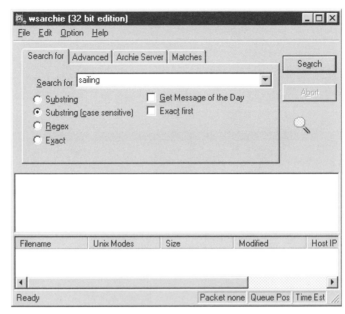

enormous reservoir. At this point, Archie, a small program that assists in the search for information on FTP servers can help you. Archie currently contains details on over 1,400 archives which can be reached via FTP. Millions of pieces of data are recorded here, which are stored on the Internet. Taken as a whole, this represents a database of countless gigabytes. Day by day, more files are added. There are **Archie servers** around the world in many countries. They contain a list of all the files that can be reached via FTP. Every month, this list is updated and all newly added data is included. Using Archie is simple: you simply enter the name of a file and Archie rummages through the Internet for this, until the program arrives at a positive or negative result.

To search for a particular file on an FTP server, you will need special software. There are thousands of Archie programs on the Internet or on CD-ROMs that are attached to numerous **specialist publications** free of charge.

41

Gopher and WAIS

Gopher was developed in 1991 by the University of Minnesota with the aim of structuring access to the various resources on the Internet as simply and transparently as possible. Gopher represents a tool for **leafing** through large volumes of information and can be most closely compared to a menu system for the Internet, via which you can get at the information sought in a similar way to using a table of

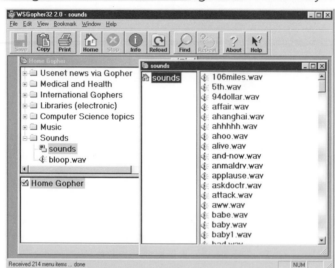

contents. Hence the name Gopher (a type of mole) – they burrow through the Internet. To connect to a Gopher server, you need more special software which offers you these possibilities. As with all other programs, these are easy to find on the Internet, if you are looking for such a program free of charge.

Gophers have become increasingly popular as a simple method of finding information quickly.

WAIS is an abbreviation of Wide Area Information Server. This is a process in which a search is made in databases on the Internet for stored information. In combination with Gopher, WAIS can be used very effectively and successfully in most cases. If Gopher is a table of contents, WAIS is the corresponding index. Gophers and WAIS are mostly used together, to access the desired information as quickly as possible.

Search engines on the WWW

Due to the enormous popularity of the World Wide Web, the search engines found there are also extremely popular – and for good reason. The World Wide Web is the Internet service which has made the Internet so famous. This new medium is also growing accordingly. Thousands of new web servers are connected to the Internet every month. This is reason enough to mention at this point the assistance available to the Internet traveller in finding the correct web address.

We will go into this again in greater detail later in the book when we look at the operation and functioning of a WWW search engine.

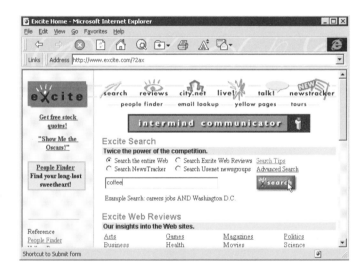

What is an online service?

Maybe, you have already heard the term **online service** in connection with the Internet. An online service can be thought of as a type of oversized data library, whose users can call up data and information and also deposit information. In contrast to the Internet, where all data is spread right across the globe, an online service combines all information at one central point. In the same way as on the Internet, communication between your PC and the supercomputers of the online service is via the telephone line.

43

Besides the enormous variety of the Internet, you also have access to its special services in an online service. You often pay a considerable surcharge, however, for this double use. Therefore calculate precisely and decide whether this method of Internet access is also the most cost-effective.

In contrast to the Internet, an online service belongs to a company, which maintains it and manages and updates the information. The operators of an online service naturally charge for these services, since

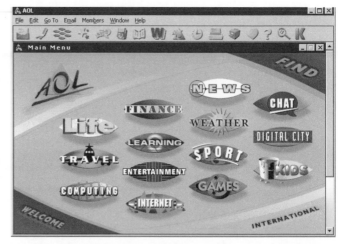

these companies want to make money. Online services include, for example, AOL, CompuServe (CSi), the Microsoft Network (MSN) and Virgin Net.

Since the Internet has become so popular, more and more online services are opening their doors to the Internet. Members of an online service therefore have the possibility of accessing the Internet from this service.

Online services have responded to the online boom in recent months and years and now offer customer-friendly tariffs, which represent a serious alternative to 'proper' **Internet access**.
The term mailbox often crops up in discussions of the Internet and

online services. A **mailbox** is a legacy of the old days when online services were still in short trousers and the Internet was only something for academics and weirdos. Like an online service, a mailbox is a way in which information can be called up via the telephone line and computer, but with one big difference. Whilst the sphere of activity of a mailbox is generally limited by region and the operator mostly offers connection nodes only at one point, an online service is available to a far larger community – nationally or even internationally.

What's in this chapter?

The following pages will tell you the requirements from your computer for active participation on the Internet. You will learn everything that needs to be known about modems and software, so that your stay on the Internet can be a pleasurable one and does not lead to frustration from the outset. We will learn about the different types of modem and take a look at the options built into Windows 95 which give access to the Internet.

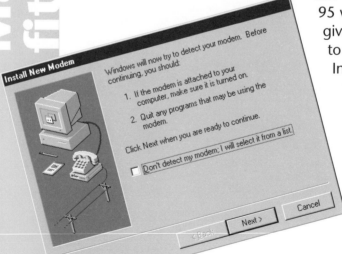

You already know:

You are going to learn:

Modem & Co.

To understand how a computer makes contact with the Internet, some basic technical knowledge is required. But don't panic. You don't have to be a 'hardware geek' or 'software junky' to follow this. If you are thinking of expensive specialist equipment, expensive high-performance computers with powerful processors and complicated adaptation work, forget it and be thankful for new technology. If you had had this book ten years ago, for example, this chapter would have filled the entire book. The technical requirements that were essential then demanded a course of study from the person concerned. Times have changed, however, and seriously so! **Computers** filled whole rooms in those days and were very difficult to operate. PCs are today available in almost every department store and much more cheaply. For anyone who still finds this purchase too expensive you can pick up a cheap computer at countless computer flea markets. To be a 'data traveller' on the Internet does not require an all-powerful computer costing thousands of pounds. You don't need an 'Arabian thoroughbred' but a solid 'workhorse'. But do check that the corresponding operating system, in our case Windows 95, is running smoothly and without errors. If it is, nothing more can actually go wrong. We explain below firstly, what hardware and software requirements have to be fulfilled to give you painless access to the Internet:

The computer is our number-one hardware component. Forget

particular brands and companies. The main thing is the **operating system**. In our workshop, we are referring to Windows 95, since this has become as widespread as the computer itself. Naturally, it is also possible to conquer the

Internet with other operating systems such as OS/2, an Apple Macintosh, an Amiga and so on.

A glance at the telephone socket

The basic requirement for your computer to be able to connect to the Internet is a telephone connection, as exists in millions of households. In this case, a **double connection** would be an advantage, so that you could continue to telephone as before and still be reached, whilst using the second line to surf. This optimum set-up is not essential, however, and a simple telephone connection serves our purposes. You do need to have a telephone socket somewhere close to your computer so that your modem can be plugged into it.

On the subject of telephones, if you have the Call Waiting service on your line, make sure you switch it off every time you go online (by dialling # 43 #) and back on when you have finished (* 43 #). Otherwise, if someone rings you while you are surfing, the incoming call will disconnect your from the Net and whatever you were doing online will be cancelled.

49

The modem is your gateway to the world

Now it's time to meet the first high-tech device you need to access the Internet – the **modem**. The modem connects your PC to the telephone network. You therefore have the possibility of transmitting programs and text over the telephone line to other computer users and, conversely, receiving data. It can also be said that a modem acts as an **interpreter** between the language of the PC and that of the telephone, so that these two worlds can talk to one another.

What are we actually doing when we telephone? Quite simply, we are transmitting sounds, in order to be able to hold a conversation with the party at the other end of the line. What appears so easy and simple at this point is a major problem for our PC, however. Computers store their information digitally, in the form of ones and zeroes. Speech is transmitted in analogue form, however, down the telephone line with electrical **oscillations**. This means that for your PC, that its data sequences must first be converted into suitable sound signals, before it can send them over the telephone line around the world. Moreover, the incoming sounds received must be converted back into the corresponding data that your computer understands. The modem is responsible for this conversion. It is the 'broker' between these two processes. 'Modem' is a word made up from the combination of '**mo**dulator' and '**dem**odulator', since the modem changes (modulates) outgoing data when sending the electrical signals on the telephone line and decodes (demodulates) the incoming signals when receiving.

How fast should the modem be?

There are as many types of modem as there are stars in the sky and choosing one is no easy matter. The main decision criterion in buying a modem is certainly the speed. Generally, it can be said: the faster a modem can 'drive' data over the

A fast modem pays for itself. If you buy a new modem, therefore, you should buy one with a speed of at least 28,800 baud. This costs somewhat more but quickly pays for itself, since you have significantly shorter downloading times from the Internet.

WHAT'S THIS?

Bit: A bit is the smallest unit of information that a PC can process. A bit can have the valu zero or one.

telephone line, the more fun there is surfing the Internet. The speed of a modem is given in 'baud'. One **baud** corresponds to one bit per second.

You will also sometimes encounter the term 'bps' in magazines and specialist shops, which means exactly the same as baud, i.e. 'bits per second'.

A computer requires eight bits, in order to store one character (a letter or digit). To find out how many characters a modem transmits in one second, divide its transmission rate by 8. For example, a 14,400 baud modem sends or receives 1,800 characters per second (14,400 ÷ 8 = 1,800). Modems with speeds below 14,400 baud should be avoided, since surfing the Internet otherwise turns into an exercise in patience. As the prices for modems have continued to fall in recent months and years, you can today buy a modem with a speed of 28,800 baud for under £200. The latest development are modems with a speed of **56,000 baud**. They are the Ferraris of the modem world.

But don't get too excited! The transmission speeds calculated above are only maximum values, which are rarely achieved in practice. The blame for this often lies in 'poor' telephone connections, with the crackling and hissing. The transmission rates are also lower because in addition to the basic data, feedback is also sent, on the basis of which senders and recipients can check whether the data has arrived correctly.

What does 'Hayes-compatible' mean?

When you have decided to buy a modem, you must definitely check that the modem concerned is 'Hayes-compatible'. The American company Hayes was a pioneer in modem development. Hayes developed a set of commands with which your computer and modem can 'talk' to one another. There are commands that regulate the transmission speed or tell the modem how it should act if it hears an engaged tone. Because all these commands begin with the letters 'AT', the Hayes command set is also termed the **AT standard**.

Over the course of time, the command set developed into an international standard. If a modem complies with this AT standard, it is Hayes-compatible. The great majority of data telecommunications software offered on the enormous market relies on the fact that your modem is Hayes-compatible. If this is not the case, there are considerable problems in the field of communication and the corresponding software falls into severe difficulties. The majority of modems available, however, are Hayes-compatible so you should not have problems in this area. Even so, when buying a modem always ask whether it is Hayes-compatible.

Does it have to be a fax-modem?

Many of the modems currently available no longer have only one section, which can transmit data, but are also able to send and receive **faxes**. These modems contain the full electronics of a fax machine, as you perhaps already know from various offices. Whilst even cheap fax machines still cost £200, you pay less for a good 14,400 or 28,800 fax-modem. Fax modems convert your PC into a fax machine, which compares very favourably with conventional fax machines – and is often better. Thanks to intelligently designed **fax software**, you can send faxes much more easily than on a conventional

For the sake of simplicity, combined data/fax modems are generally termed fax modems. Pure fax modems are not generally available!

fax machine.

If you have the chance to buy such a modem, do so. Even if you currently do not have the need for a combined machine, the time will come when you will appreciate the advantages of a fax modem.

The voice modem

For some time, a new type of modem has been gaining popularity. These 'voice modems' are capable of transporting voice data at the same time as a data or fax transmission running in the background. In this way, for example, you can talk to a subscriber by telephone and send a file, without having to interrupt the conversation.

It's faster with ISDN

You will experience the real thrill of speed if you have access to the ISDN network via a digital telephone connection. **ISDN** enables you to reach far higher transmission speeds than with analogue modems. Using ISDN, you can access the Internet at a speed of 64,000 bps (compared to the fastest analogue modems at 'only' 56,000 bps).

Speed comes at a price, however. ISDN is worthwhile only if you either already have an ISDN connection or need to download masses of data every day from the Internet. In that case, an expensive ISDN connection is worthwhile.

If you have opted for an ISDN connection and have had this installed by British Telecom, it is time to decide on an ISDN PC card. The range here is divided into **active** and **passive** cards. Active ISDN cards contain their own processor, which takes on most activities during a data transmission and significantly alleviates the burden on the computer's processor. The disadvantage is that such cards are considerably more expensive than their passive counterparts.

Passive cards do not have their own processor but use the computer's processor for data transmission activities. If you intend to transmit large volumes of data, for example to hold video conferences, then an active ISDN card is recommended.

The installation of an ISDN card is carried out in the same way as for a modem (we will see how to install a modem later in this chapter). Generally, ISDN cards are provided with an installation program, which makes the installation process an almost automatic one.

53

Without software, it won't work

The basis for our excursions onto the Internet is Windows 95. With this operating system from Microsoft, you have the opportunity, for the first time, to make full use of communication. Windows 95 is a graphical **operating system**, which you can operate with the mouse. So what is an operating system? Very simply, the operating system brings the 'dead' hardware, i.e. your computer, to 'life'. An operating system, to a certain extent represents a type of interpreter, which translates all the entries you make through Windows (mouse clicks, program call-ups, etc.) into ones and zeroes, which is all your computer can understand. Windows 95 is the most widespread graphical operating system across the world, successfully superseding the earlier Windows 3.x. The sustained success of **Windows 95** lies in the fact that you simply switch on your computer and begin without any great computing knowledge.

Windows 95 and your modem

Once you are sure that the serial interface of your computer is fully operational, your modem can be connected into the operating system. This step is necessary, since you have to tell Windows 95 what modem is connected to your computer. A useful device known as 'modem wizard' helps you install a modem on Windows 95 and takes over most of the installation work.

To set up a new modem, you must first call up **Control Panel** from Windows 95. This is the first step for installing modems on Windows 95. Under Control Panel, you have the possibility of making Windows 95 aware of all hardware and software components. We will use this possibility to 'acquaint' our modem with Windows 95.

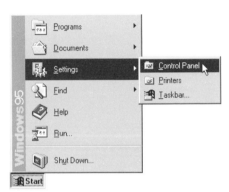

Click on the *Start* button. Put the mouse pointer on the menu item SETTINGS and click on CONTROL PANEL in the submenu.

Double click on the *Modems* symbol.

55

Now you have arrived at the **Modem Wizard**. As you can see, this is blank. This should not come as any surprise, since there is no modem installed as yet. To make the new modem known, we must now add it.

After clicking on the *Add* button, a dialogue field appears, in which you must inform the modem wizard of what type of modem you wish to install. You are given the choice of a PCMCIA modem card for notebooks and *Other*. We opt for *Other*, since we are assuming that this is not a notebook.

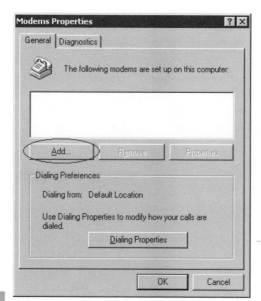

1 Click on the *Add* button in the *Modem Properties* dialogue field.

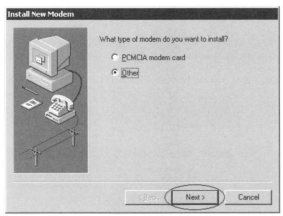

2 Click on the
Other option and then
on the *Next* button.

Now it gets serious. Deactivate the checkbox *Don't detect my modem; I will select it from a list* in the dialogue field displayed next. Windows 95 then tries to identify devices connected at the individual interfaces automatically (in our case, the modem). It is important for this that the modem is also switched on (if it is an external modem). If you activate the checkbox, Windows 95 leaves it to you to enter the modem manually. This option should be considered only for 'exotic' modems.

A computer communicates with external devices via **interfaces**, such as printers, modems, scanners and cameras.

Windows 95 examines the **interfaces** and checks whether a modem can be found there. Have a little patience at this point, as this process can take a few minutes. So as not to panic if the computer crashes (which can, of course, also happen), you must look at the external modem. A couple of lights should illuminate here periodically. This is a sure sign that your computer is still active and still in the land of the living.

57

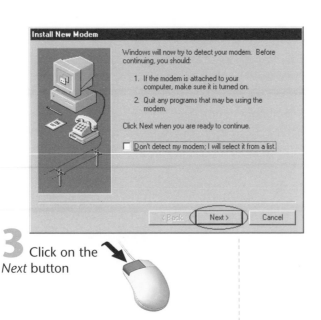

3 Click on the *Next* button

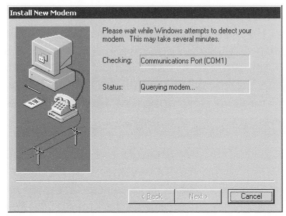

4 The interfaces are then polled to find a modem – all you do now is watch and wait.

WITHOUT SOFTWARE, IT WON'T WORK

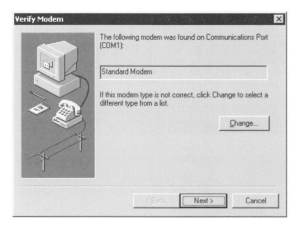

If the search process was successfully completed, the window shown here appears. In that case, you can breathe easily; the computer and Windows 95 have detected your modem. If you are using a popular modem known to Microsoft, the correct type appears at this point and you do not need to bother about any further configuration tasks. Otherwise (as in our case), Windows 95 suggests a **standard modem**. This setting generally suits all modems. By clicking on the *Change* button, you have the possibility of selecting your modem type from a list.

In the list field of the window now appearing, you see a selection of companies which manufacture modems. Generally, your **modem manufacturer** should be represented here. Once you have found it on the list, click on the company, using the mouse, and in the right-hand list field you will see a list with the modems of the corresponding manufacturer. If this is not the case, there is still the possibility of copying the necessary files using the button *Have Disk*. In most modem packages, there is a disk with the relevant **drivers** for Windows 95 – this disk is now requested.

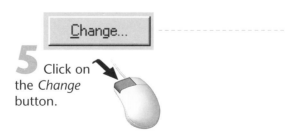

5 Click on the *Change* button.

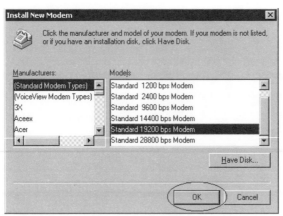

6 Select your modem and then click on the *OK* button.

After you have selected your modem, Windows 95 shows you the modem type and relevant interface again. Is everything correct? If not, then do not continue.

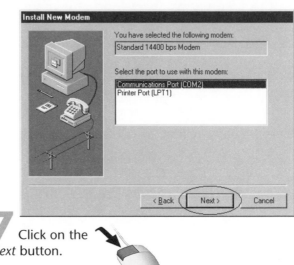

7 Click on the *Next* button.

8 Click on the
Finish button.

There you have it! Your modem is now fully installed and configured.
Nothing should now stand in the way of your surfing.

Installing the TCP/IP protocol

As you already know, we need a protocol to be able to receive and send any data via the Internet. The name of this protocol is **TCP/IP**. This transmission protocol is used by the entire Internet community as the generally applicable transmission standard. In Windows 95, we must explicitly install and set up this protocol.

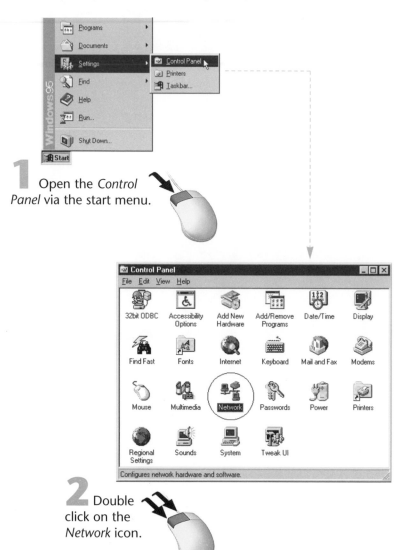

Open the *Control Panel* via the start menu.

Double click on the *Network* icon.

After you have called up the **network configuration** window under Windows 95, you may already have some entries in this window, as in our case. There are **network protocols** and other devices used to attach your computer to a network. However, on your computer, this window may be empty.

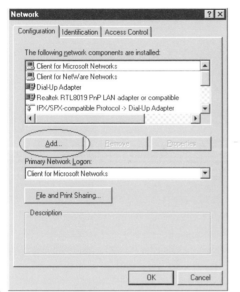

We will next set up the Microsoft Dial-Up Adapter and the TCP/IP protocol, so that your computer can communicate with the Internet.

3 Click on the *Add* button.

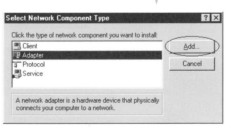

4 Double-clicklick on the *Adapter* entry (even if your do not have one, Windows 95 also considers modems to be an adapter) and then click on the *Add* button.

63

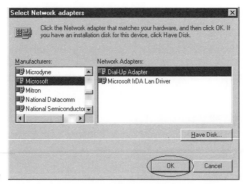

5 Select the *Microsoft* entry and highlight
Dial-Up Adapter in the right-hand list.
Confirm your choice with the OK button.

In the next installation stage, we will set up the actual protocol, in
our case TCP/IP, and select the Microsoft TCP/IP protocol from the
Microsoft folder. Then, you should see your two newly installed
components (**Dial-Up Adapter** and TCP/IP) in the network
configuration window. If this is the case, click with the mouse on
OK. Windows 95 then asks you to insert the Windows 95
installation CD, so that the relevant files can be copied. Once this
has been completed, you have to reboot the computer, to activate
all the changes.

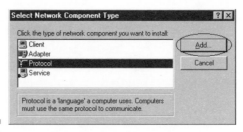

6 Select the *Protocol* entry
and then click on *Add*.

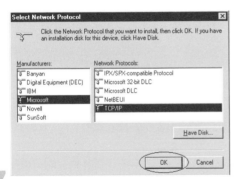

7 Highlight the *Microsoft* entry and then select the *TCP/IP* protocol on the right-hand side. Confirm your choice with *OK*.

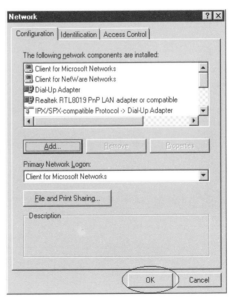

8 Dial-Up adapter and TCP/IP are now listed in the dialogue field. Close this by clicking on *OK*.

65

The path to the Internet

Now that we have installed our modem and TCP/IP, it is time to make contact with the Internet. The Windows **Internet Connection Wizard** helps you establish a connection with an Internet service provider. This program asks for all the necessary details in sequence and then makes all the necessary changes for you.

Protect your **password** against unauthorised access. Anyone who knows your password or code can pass themselves off as you on the Internet and surf at your expense.

You should perform these installation steps only if you have received all the necessary details and documents (such as telephone number, password, etc.) from your Internet service provider with whom you need to make a connection and you are ready to set up your computer to connect to the Internet.

Make sure you have the Windows 95 installation CD-ROM or diskettes to hand (Windows will need to copy some files to your system during these steps).

1 Select the GET ON THE INTERNET menu item.

Once you have activated the relevant menu item, you are welcomed by the Internet Connection Wizard. By clicking on the *Next* button, you reach the window in which you can specify the installation process.

You have the choice at this point between an automatic, manual and current **installation routine**. We will select the manual installation routine, in order to give you a detailed insight into the functioning of the Windows 95 Internet connection.

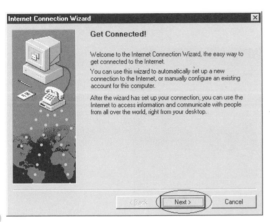

2 Click on the *Next* button.

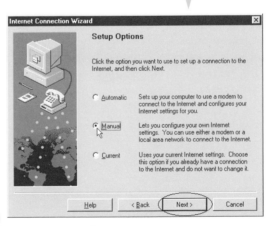

3 Select the *Manual* option and confirm with *Next*.

67

Now we get down to business. For the first time, we are officially greeted by the Internet Connection Wizard. We can safely leave this window quickly, to reach the point that establishes the way in which you connect to the Internet. The option *Connect using my phone line* is the one to choose. This means that you have a stand-alone PC and a **modem** and intend to use these to make a connection to the Internet. The option *Connect using my Local Area Network* does not interest us, since this field is generally activated only if your computer is in a network. In that case the computers make a connection with the Internet via the **network** connection. Since this is rarely the case in the home, we shall ignore this option.

4 Welcome to the Internet Connection Wizard. Simply click on the *Next* button.

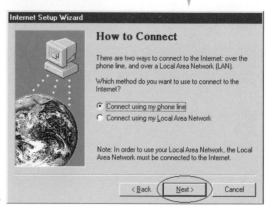

5 Click on the *Connect using my phone line* option field and then on *Next*.

Now that we have established that we need a modem (which we have already installed) to make contact with the Internet, we must first give the connection profile a name. At this point, a **profile** describes all the settings (telephone number, login name, password, etc.) that are necessary to make contact with the Internet service provider. Once this has taken place, we enter the telephone number, which we wish to use to dial onto the Internet. These telephone numbers can be obtained from the documents of the Internet service provider.

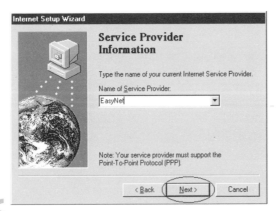

6 Give your Internet profile a meaningful name. Confirm your choice with *Next*.

69

Enter the dial code of the Internet service provider in the *Area Code* field and the telephone number in the input field of the same name. In the *Country Code* field, you must enter the country code – in our case United Kingdom. Confirm these entries by clicking the *Next* button.

When all the necessary details have been given and your computer and modem know where they should dial onto the Internet, it is time to enter your **login name** and **password**. You will find these details in the documents sent to you by your Internet service provider.

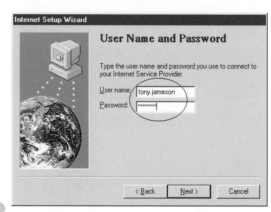

Enter your name in the *User* name field and your secret password under *Password*. Complete all entries using *Next*.

After you have entered the **user name** and **password**, the Internet Connection Wizard asks you to enter an IP number. As you have already discovered, every computer on the Internet is uniquely identified using this number. If the Internet service provider has assigned you a fixed IP address, select the option *Always use the following*. If the service provider assigns you a dynamic IP, select the *My Internet service provider automatically assigns me one* option. There is an idea behind this. Every Internet service provider has a large pool of IP addresses, which it can issue freely. When you dial an Internet service provider, it assigns you a temporary address dynamically. Let us assume that the service provider assigns you an IP address every time, as this is most likely to be the case.

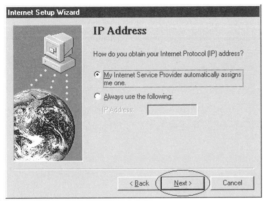

9 Select the option *My Internet service provider automatically assigns me one* and then click on *Next*.

By clicking on the *Next* button you reach a window in which you need to enter the IP address of the **DNS server** and (if your Internet service provider gave you details of a secondary DNS server) that of an alternate DNS server in corresponding input fields. These will be 'number-and-dots' addresses that look something like those in the following screenshot:

71

10 Enter the IP address of the DNS server into the field of the same name and (if present) that of the second DNS server. Click on the *Next* button.

We have almost finished. All entries have now been made and the Internet Connection Wizard can now copy all files and transfer all entries into the relevant files.

After the installation and configuration have been completed, you must reboot your computer, so that all **new settings** are activated.

Once the computer has restarted, you can make a double click on the *My Computer* icon on Desktop. In the following window, there is a program group with the name **Dial-Up Networking**. Here you will find your newly produced profile, which will help you to contact the Internet service provider.

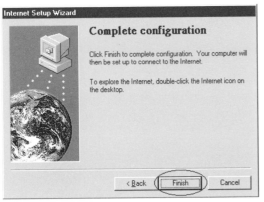

11 Click on
the *Finish* button.

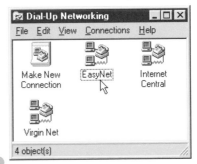

12 There it is, your newly
established connection to the Internet.

Onto the Internet

Now that we have installed all that is needed to dial onto the
Internet or the Internet service provider, we can get on and do it.

Open the *My Computer* program group to click on the Dial-Up
Network icon.

The window opens, showing all the computer's modem
connections – for example our connection to the Internet service
provider EasyNet.

73

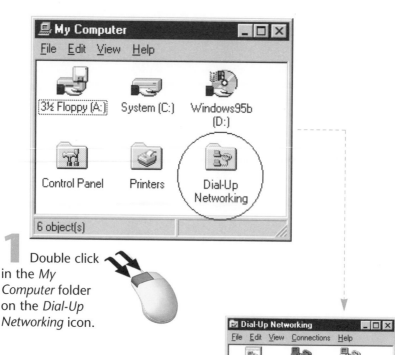

1 Double click in the *My Computer* folder on the *Dial-Up Networking* icon.

2 Click with the right-hand mouse button on the newly created connection with the Internet service provider.

To make a connection to the Internet service provider, you must click, using the right-hand mouse button, on the relevant icon. In this way, the **properties** of the icon concerned are displayed. The first menu item offered is CONNECT. Click on this menu item.

Thereupon, a window appears, showing all the data that has already been entered in the framework of installing our profile.

This includes the user name and password. You can still change these two pieces of information, if something has changed. If you activate the *Save password* checkbox, you no longer have to enter your password explicitly in future.

As soon as you click on *Connect*, your computer and the modem make the connection with the Internet service provider and establish a permanent connection, which you can end again by clicking on the *Disconnect* button.

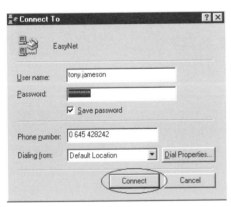

1 Click on the *Connect* button

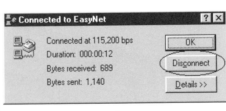

2 A mouse click on the *Disconnect* button ends the connection to the Internet.

What's in this chapter?

Now that all the requirements to connect to the Internet have been established in the previous chapters, it is time to install a WWW browser. This will give you access to one of the most popular Internet services – surfing! You also learn how to handle the WWW browser, so that you can delve into WWW after just a few moments. Microsoft Explorer which is contained in Windows 95 as standard, can be used as a web browser.

You already know:

You are going to learn:

A **web browser** is a program with the aid of which you can surf around the World Wide Web.

Installing a web browser

The web browser is the essential tool to be able to exploit the enormous range of information on the World Wide Web. Many companies in the past set about developing web browsers. Two alone were able to divide up the market between them, however, and gain the favour of the Internet community – the American companies Microsoft and **Netscape**. In the framework of this workshop, we will install the **Microsoft Internet Explorer 3.0** web browser. You can obtain the latest Explorer software either from your Internet service provider, a software dealer, as a copy from a friend (that is allowed at this point) or directly from the Internet. Another rich hunting ground are specialist computer magazines, which regularly include a CD ROM that has many useful programs.

Installing the compressed version

Most programs that you obtain via the Internet or from CD ROMs are in a packed file format. This simply means that the original file was compressed to a minimum of its original size, so as to be able to transmit it faster over the Internet. In order to unpack these files again, a program is required that restores the program to its original size after a successful **download**. There are heaps of programs of this type on the Internet or on CD ROMs from computer magazines. There are, however, compressed files with the file ending .EXE, which unpack themselves without the need for a special utility – as is the case for our Microsoft Explorer 3.0 installation. In our example, I have obtained the Explorer from a CD ROM from a magazine.

Firstly, the installation must run. For this purpose, click on the *Start* button of the Windows 95 interface and then on *Run*. Once this is done, enter the name of the compressed file or search for it. By pressing the *Browse* button, you can gradually move through the structure of your hard disk.

INSTALLING A WEB BROWSER

Next confirm the path to the file with *OK*, so that the computer can start to decompress the file. As a backup, you are asked again beforehand whether you would actually like to install the Microsoft Internet Explorer.

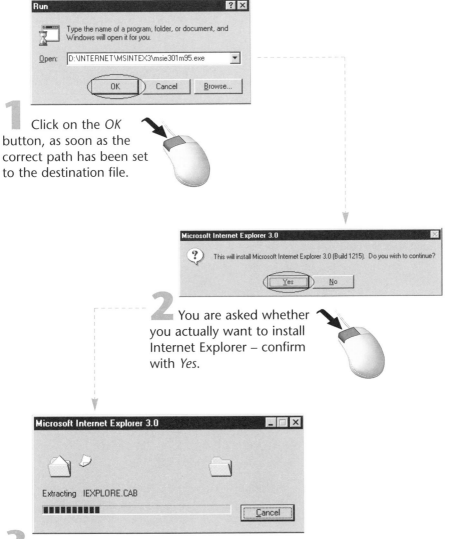

1 Click on the *OK* button, as soon as the correct path has been set to the destination file.

2 You are asked whether you actually want to install Internet Explorer – confirm with *Yes*.

3 The installation routine unpacks the compressed file into its individual components.

Once all the files have been decompressed from the original file, the installation itself follows. You are asked whether you accept the general Microsoft **licence conditions** affecting Internet Explorer. Carefully read through the licence contract and then click on the *I Agree* button.

The installation application then begins to copy the corresponding files into the destination folder.

4 Click on *Agree*, if you have no objection to the Microsoft licence agreement.

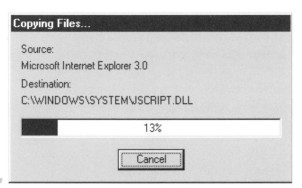

5 The installation routine copies all files into the corresponding destination folder.

Under certain circumstances, there may already be some files on your hard disk which are of a more recent date. In that case, you should always keep the more recent file. Once all files have been successfully installed, the computer asks whether it should restart to activate all the changes.

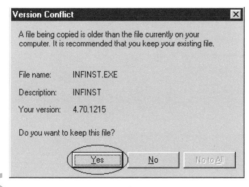

6 If possible, always keep all your more recent files and confirm with *Yes* .

81

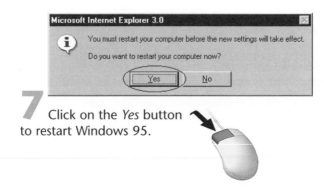

7 Click on the *Yes* button to restart Windows 95.

After a **restart**, an icon appears on the Windows desktop with the caption *The Internet*. This is used to start Internet Explorer.

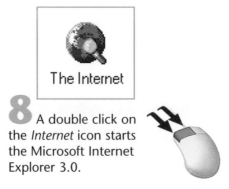

The Internet

8 A double click on the *Internet* icon starts the Microsoft Internet Explorer 3.0.

Microsoft Internet Explorer 3.0 in detail

Before you can properly use Internet Explorer, some basic settings are needed, to adjust the web browser to your circumstances. These include, for example, the e-mail address setting and the definition of the corresponding Internet profile.

Now start Internet Explorer by double clicking on the *Internet* icon on your Windows 95 interface. The Explorer then opens initially with an error message, since the web browser is attempting to access a particular web address – in this case, that of Microsoft, since it is set as a default. Firstly, you must tell Internet Explorer what profile it should use to make contact with the Internet service provider (the profile that we created in the previous chapter).

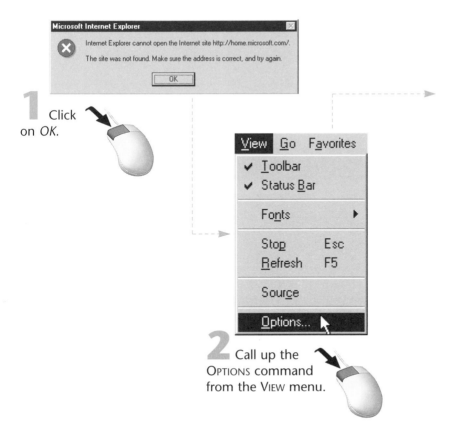

1 Click on *OK*.

2 Call up the OPTIONS command from the VIEW menu.

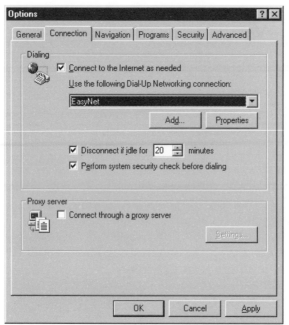

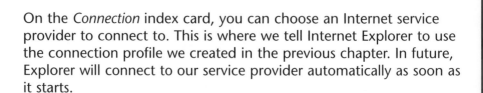

3 Select the
Connection tab.

On the *Connection* index card, you can choose an Internet service provider to connect to. This is where we tell Internet Explorer to use the connection profile we created in the previous chapter. In future, Explorer will connect to our service provider automatically as soon as it starts.

At this point, you can also adjust the time setting, which terminates the connection after a particular period of inactivity. This means that if you leave your computer and forget to disconnect from the Net, Internet Explorer will do it for you after a few minutes to save you from running up big bills. To do this, you simply have to activate the checkbox *Disconnect if idle for* and specify the desired period in the next field.

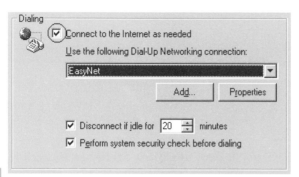

4 Activate the checkbox *Connect to the Internet as needed*. Next, click on the list field below and select the connection profile you created earlier.

As you have perhaps noticed, there was a checkbox on the index card with the phrase *Connect through a proxy server* – what does this mean?

Proxy servers are computers that buffer-store various pages from the World Wide Web locally at the Internet service provider. If these are later requested again, they can be loaded directly from there. This has the advantage that these pages do not have to be first downloaded from the Internet in a time-consuming fashion. Access to frequently required WWW pages is therefore greatly accelerated.

There are, however, possible disadvantages in using a proxy server. Since all communication with the Internet flows through the proxy server, there is the possibility of someone **monitoring** and judging your surfing behaviour. After all, which Internet sites you are drawn to is no one else's business. Moreover, proxy servers are updated only at particular intervals, so you may be working with pages on a proxy server that are already several days old, whilst the original pages have long since been updated on the Internet.

Ignore this option, since the Internet has now become faster. At this point you would only have to deal with the negative sides of the proxy server.

85

Finally, to complete the adjustment work, enter the size of the temporary Internet folder on your hard disk. Internet Explorer deposits here all graphics and documents that were viewed in the most recent Internet tours. This buffer memory, which is also termed a **cache**, accelerates surfing on the World Wide Web, since the graphics are then partly already on the local hard disk. From time to time, however, this folder must be emptied, in order not to fill your entire hard disk with 'Internet garbage'. To minimise this work, Internet Explorer offers a corresponding automatic mechanism.

WHAT'S THIS?

Cache is a particular memory in your computer, which has the task of holding frequently required files for the processor and providing these as required.

Activate the *Advanced* tab in the *Options* window. We need to look at the *Temporary Internet Files* section.

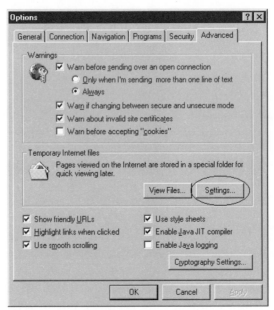

1 Click on the *Settings* button.

If you click on the slide gauge and move back and forth, keeping the mouse button depressed, you can change the size of the memory. Here, Internet Explorer always uses a percentage of the absolute hard disk size. For example, if you have a 1 gigabyte hard disk and the size of the temporary Internet files is set to 4 per cent, Internet Explorer will use a maximum of 4 megabytes for this folder.

By clicking on the *Empty Folder* button, all temporary Internet files are deleted.

2 Click on the *OK* button, to end the entries in question.

Initial excursions on the World Wide Web

You can imagine the World Wide Web as an oversized newsagent, which spans the entire world. The services offered in the framework of this Internet service are structured in a similar way to magazines – they consist of pages you can browse through with the aid of your Internet Explorer.

So that the web browser knows from where to fetch the desired pages or contents, it needs a destination address. This appears as follows: *http://home.microsoft.com*.

You have probably seen a line such as this before in a newspaper or on television or heard it on the radio. But what do all these abbreviations mean? All the points, slashes and abbreviations are a 'Uniform Resource Locator' or URL. A URL is simply the precise address of a document or web server on the Internet.

Since there is no table of contents of all the web pages on the Internet, the inventors of the process used URL as a way of describing a path (locator) to the destination (resource) and in such a way that these appear the same all over the world (uniform).

The 'http://' at the start of the address means that all data after the **Hypertext Transfer Protocol** is to be transferred. This implies a standard process by which pages from the World Wide Web (i.e. the hypertext of a web page) is fetched onto your computer. You already know what a protocol is from the previous chapters, but let us remind ourselves. A protocol regulates the transport of data via a cable medium (in our case the Internet), so that all data arrives correctly at the recipient and nothing is lost along the way. In the case of data loss, a transmission protocol also describes in detail how to proceed. But, of course, there's no need to worry about that. All the functions of a transmission protocol run invisibly in the background.

Let's take an example of a URL and look at its constituent parts:

http://www.lon.ac.uk/public/fun/index.htm

The *http://* prefix tells Internet Explorer that the HTTP protocol should be used, indicating that this is a file on the World Wide Web.

The specification .uk tells us that this WWW server is in the United Kingdom.

The *ac* abbreviation means that this is a university or college, and *lon* is an abbreviation for London, so this is the London University computer.

☞ Once it is connected to the server, Internet Explorer looks in a directory it finds there called *public*.

☞ Inside the 'public' directory is a subdirectory called *fun*.

☞ The document concerned is in the 'fun' directory on the computer.

☞ Finally, the document *index.htm* can be accessed.

Setting up a new start page

So, after you have learned a little theory about addressing on the World Wide Web, it is time to put this knowledge into practice and gain a feel for Internet Explorer.

Start Internet Explorer from the Windows 95 interface, by double clicking on the *Internet* icon. There then appears a window, which connects you with the Internet service provider. Once this connection has been successfully established, Internet Explorer begins with the predefined **start page** in the Web browser. At the moment, by default, this is the Microsoft web site.

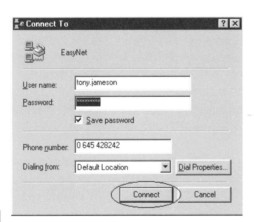

After entering the password, click on the *Connect* button.

Connected to EasyNet   ? X

Connected at 115,200 bps
Duration: 000:00:12
Bytes received: 689
Bytes sent: 1,140

OK

Dis_connect

_Details >>

2 A connection is established with the Internet service provider.

3 The predefined start page is loaded.

During the transmission of data, the Explorer symbol moves in the top right-hand corner of the program window. This has the advantage that you always know whether Internet Explorer is active or inactive.

As you can see, the Microsoft page is loaded first, which is set as a default. In the next step, let's change Internet Explorer's start page.

Find another page you would like to be greeted by. As a keen cinemagoer, I would prefer Internet Explorer to display my favourite movie site when it starts *http://movieweb.com.*

You also have the option to load only parts of a web page. If the loading process takes too long for you, which can often be the case with large graphics, you can break off transmission using the Stop button. All data received up to this point is then displayed to you.

For this purpose, call up the OPTIONS command from the VIEW menu. In the dialogue window then appearing, activate the index card with the title NAVIGATION. Next click on the USE CURRENT button.

4 Click on the *Current Page* button and then enter the URL (in our case, *http://movieweb.com*) in the *Address* field.

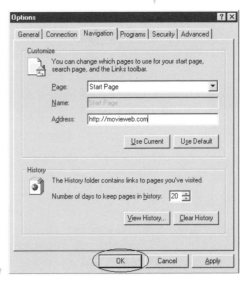

5 Close your entries with *OK* and restart Internet Explorer.

6 There you have it, your new start page.

Producing a bookmark list

A good question at this point is, After typing a URL into Explorer's address box, how can you save yourself having to retype it every time you want to visit that site again? Every web browser offers the option to create **bookmarks**. With the aid of these bookmarks, you can save the URLs of sites that you visit most frequently for quick recall. In this way, you can produce your own list of favourites on the web.

Start Internet Explorer again and connect to the Internet (you already know how to do this now). First, you are going to enter a new address, and then include this in the bookmark list and produce a new bookmark subject group for the entry concerned. After the start page has been loaded, click on the *Address* field. Internet Explorer highlights the current address in blue and you can now type a new address. I have entered the following address: *http://www.whatson.com*.

On this web page, you will find news and reviews of the latest plays, shows and musicals all around the UK. Type the URL into Internet Explorer's address field (you can leave out the 'http://' at this point).

1 Enter the new URL in the *Address* field and confirm the entry using the ⏎ key. The new page will be loaded.

You will now include this page in your bookmark list, so that it is always available.

2 Click on the *Favorites* button. Then select the command *Add To Favorites*.

93

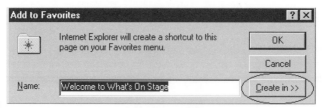

3 Give the web site a new name or leave it with the old name. Click on the *Create in>>* button.

At this point, you have the option of producing special folders to contain the URLs which are to be included in the bookmark list. You now produce a folder with the name Entertainment.

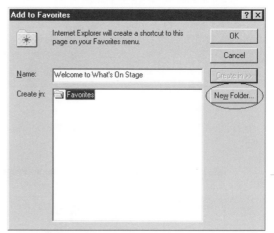

4 Click on *New Folder.*

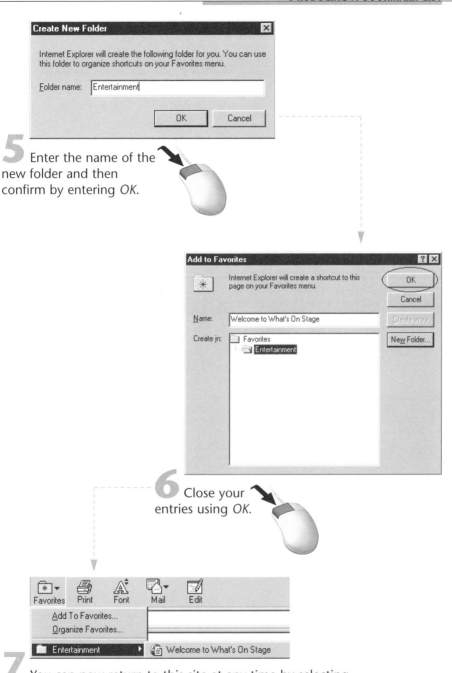

Create New Folder

Internet Explorer will create the following folder for you. You can use this folder to organize shortcuts on your Favorites menu.

Folder name: | Entertainment |

| OK | Cancel |

5 Enter the name of the new folder and then confirm by entering *OK*.

Add to Favorites

Internet Explorer will create a shortcut to this page on your Favorites menu.

OK

Cancel

Name: | Welcome to What's On Stage |

Create in: Favorites
Entertainment

Create in >>

New Folder...

6 Close your entries using *OK*.

Favorites   Print   Font   Mail   Edit

Add To Favorites...
Organize Favorites...

Entertainment   ▶   Welcome to What's On Stage

7 You can now return to this site at any time by selecting *Favorites* and clicking the entry in the new Entertainment

95

Internet Explorer

You have made your first bookmark entry and are therefore able to enter new **web addresses** on your own. To enable you to move around on a web page, you will next learn something about the Internet Explorer interface.

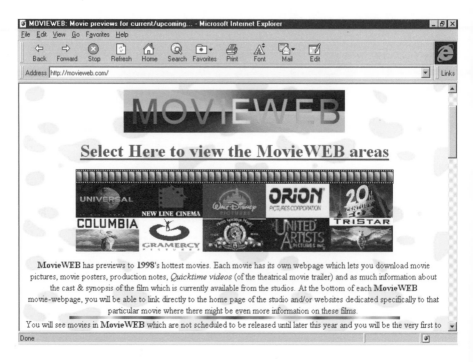

Explorer screen

The illustration above shows the screen of Internet Explorer. In principle, this is divided into three sections – you will see a menu bar, a **toolbar** and the Address field, into which you can type a URL.

To the right, alongside the toolbar, is the Internet Explorer status display. In the active state, when data is being transmitted, the 'e' changes colour – the Explorer is active. The explorer contains another bar with buttons, which cannot be seen at first. This bar is intended to support the beginner on his or her first excursions and contains links to a few web sites that Microsoft thinks you might find

useful. You reach this bar by clicking on the *Links* field below the status display.

Once you have clicked on this icon, the address field disappears and the toolbar containing the links buttons takes its place. When you need to see the Address field again, just click on the Address button at the left of the window: the Links toolbar slides back to the right and the Address field reappears.

Although they probably seems a little weird, these sliding toolbars take up less space in the window, and so leave more space for viewing the web pages themselves.

The Internet Explorer toolbar

The most important functions on Internet Explorer are in the form of a toolbar.

- *Back* returns to a previously loaded web page. If you have not loaded a page previously, this button is deactivated.

- *Forward* jumps one web page forwards. You can only use this after you have clicked the Back button at least once.

- *Stop* interrupts the transmission of a current web page from the Internet.

- *Home* loads your start page, which you have defined in Internet Explorer.

- *Search* changes to a particular web page of the Microsoft WWW server. On this page, Microsoft has combined several of the best known search engines. You can select these to arrange a search for a particular subject or term. More information on search engines is included in the next chapter 'On the hunt for information – search engines on the WWW'.

97

Favorites leads you to your bookmark entries, the addresses which you have archived for later visits.

Print passes the current web page to your printer for printing on paper.

Font gives you the option to enlarge or shrink the size of the text on a web page.

Mail offers you the option to send email messages from Internet Explorer and access the contents of newsgroups.

Saving a file onto the hard disk

Internet Explorer loads data, which you receive from the Internet in your web browser. This has the advantage that these corresponding pages are already on your computer. The option exists therefore, to save important or interesting documents on your hard disk, so that you can call these up again later, without having to return to the Internet.

There are also more possibilities. Not only can Internet Explorer display the saved documents in their **original layout**, but **hyperlinks** also operate on other documents or WWW servers.

There is a snag, however. Since pictures or graphics on a web page exist on the corresponding WWW server separately as independent files, these are not saved locally onto your hard disk when you save the web page. That means that when you open the web page you saved onto your hard disk, these image will be missing. This should not be too much of a loss, however, since the important information is mostly contained in the text.

Start your web browser as usual, by double clicking on the *Internet* icon and wait for your start page to load.

After you select the S<small>AVE AS</small> F<small>ILE</small> command, a window appears in which you can save the current web document on your hard disk.

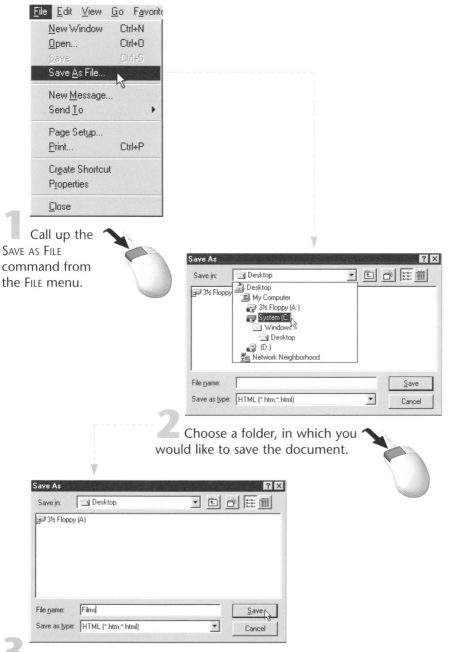

1 Call up the SAVE AS FILE command from the FILE menu.

2 Choose a folder, in which you would like to save the document.

3 Enter a name in the *File name* field, under which your document will be saved on the hard disk – in our case as 'Films' and click the *Save* button.

99

Note that the ending to the file to be saved remains *.htm* or *.html*.
So you can be sure that the file is saved in **HTML format**, i.e. in the
format that your web browser can read.

You also have the option of saving WWW documents as plain text
files. This has the advantage that the WWW pages saved in this
way can be loaded into your word processing program for further
work. To do this, simply set the file type as *Plain Text (*.txt)* in the
Save as type list.

 Click in the Save as type selection
field on the Plain Text (*.txt) entry and
then on the Save button.

Loading saved pages

The loading of web pages into the web browser is just as easy as
saving them. To do this, simply open your web browser as usual and
click on the OPEN command in the FILE menu.

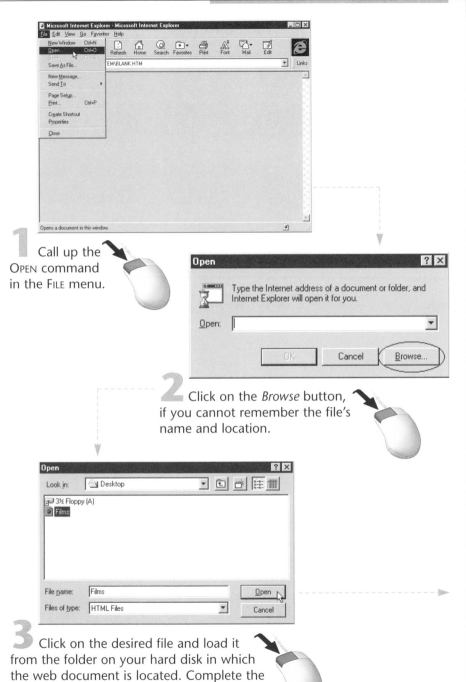

1 Call up the OPEN command in the FILE menu.

2 Click on the *Browse* button, if you cannot remember the file's name and location.

3 Click on the desired file and load it from the folder on your hard disk in which the web document is located. Complete the selection using *Open*.

101

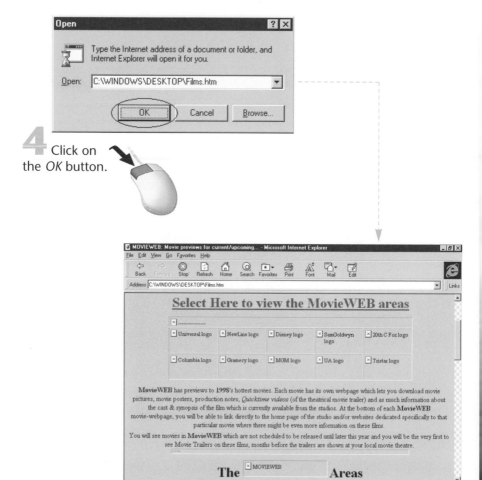

4 Click on the *OK* button.

5 The page concerned is loaded locally from your hard disk – without graphics.

Searching through WWW pages

Web pages can vary considerably in size. Some pages might contain only a couple of paragraphs of text while others hold vast amounts of information. If you are searching for a particular reference in a long page, you would rather not have to read every word until you

find it (particularly if you are online and paying for the privilege!).
For this reason, Internet Explorer includes a search facility, similar to
those found in word processors, that lets you search the current web
page for a particular word or phrase. Let's use this facility in the page
we just opened.

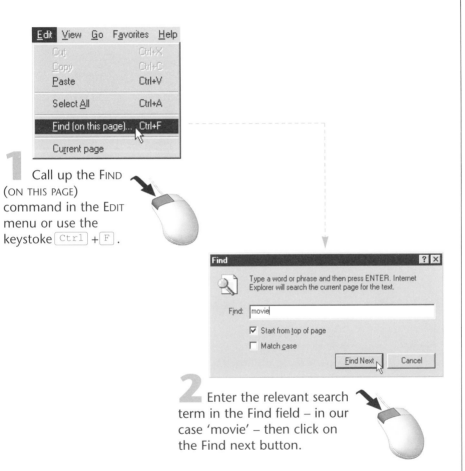

1 Call up the FIND
(ON THIS PAGE)
command in the EDIT
menu or use the
keystoke Ctrl + F .

2 Enter the relevant search
term in the Find field – in our
case 'movie' – then click on
the Find next button.

At this point, you have the option of using the checkboxes *Match
case* to refine the search process to upper- or lower-case letters.
Internet Explorer then searches through the entire document for the
term concerned. If Explorer finds the term, it halts the search and
marks the term in the document. You can then choose to continue
the search in the document using the *Find next* button.

103

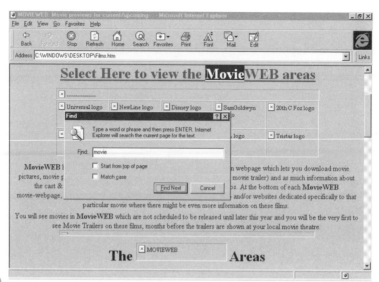

 Click on *Find next*,
to continue the search.

Printing out WWW pages

Suppose you want to **print out** a web page to show someone in hard copy form. Microsoft has thought of this. Internet Explorer offers the possibility of printing out WWW pages either as soon as you find them online or after saving them to your hard disk.

HTML is the programming language in which web pages are produced.

Start Microsoft Explorer and load a saved **HTML page** into the browser.

1 Click on the *Print* button

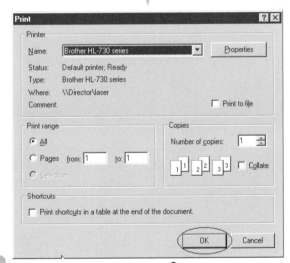

2 Click on the *OK* button and your WWW page is printed out (if you have set up the printer on Windows 95).

If you do not like the **default settings** that determine how a page is put onto paper, you can adapt these according to your **personal preferences**. Call up the PAGE SETUP command in the FILE menu.

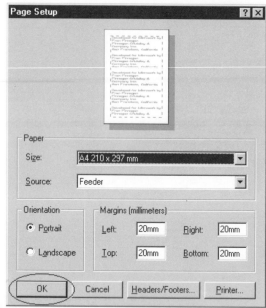

3 You can adjust page formats, margins and paper formats here. Complete your changes with a click on the *OK* button

4

What's in this chapter?

In this chapter, having mastered Microsoft
Internet Explorer in its basic form, you will
learn how to operate search engines on the
World Wide Web. We take one example and
examine it closely. You learn about the
differences between a search engine and a
web catalogue and how to use these
efficiently to meet
your objectives.
We also take a
detailed look at
the Internet
Gopher service.

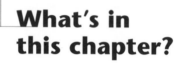

You already know:

You are going to learn:

109

How a search engine works

As you already know, the Internet is so extensive that it is becoming increasingly difficult to find the information you may be seeking. In order to get a grip on this problem, there are search tools on the World Wide Web known as **search engines**. These search engines are also termed **web crawlers** in Internet circles, because they 'crawl' through the Internet to find the desired information. Search engines are the most frequently visited sites on the Internet and generally the best starting point for any web excursion. You will also often hear the term **web spider** in connection with search engines. Spiders are programs that scour the WWW for documents, by following hyperlinks on the web pages.

A search engine frequently uses those types of spiders that search through documents on the Internet according to particular rules. As soon as a spider discovers a homepage on a specified topic, a **software agent** is appointed to fetch the URLs and documents and send information on them to a software index. The software index then receives the documents and URLs from the agents. Next, the program extracts certain information from the documents on the topic and puts this in a list – the data is 'indexed'. At this point, the various search engines differ from one another. Some index each individual word in the document. Others, on the other hand, take only the titles and the headings of the various web pages. All this information is stored in a database of the search engine.

If you are searching for information using a search engine, you generally enter key words which describe the required subject area or topic. The search engine's database is then searched using the information entered. If the search engine is successful, the results are sent back as an HTML page. At this point too, search engines differ from one another. Some rate the results to show which WWW sites correspond most closely to the search term entered.

Many display the first sentences from the document, so that you can see immediately what there is on the WWW server. If you click on the left of a document from the results list, a connection is made to this document or WWW server.

What are the differences between a search engine and a web catalogue?

An **Index** describes a list of data, which was found during a search process.

Quite simply: A catalogue system offers pages that have been assessed and sorted according to particular topics. The search for subject area or companies and organisations is simplified by the clear classification and **index**. Behind a web catalogue, there are a whole series of hard-working people. These are always searching for new web servers, which are sorted by topic into the catalogue and provided with a description. Editorial processing takes place, in order to simplify your search as much as possible. Thus a web catalogue does not search through the Internet in pursuit of the search term concerned, but draws on its enormous database of web addresses – similar to a mail order catalogue.

A search engine, by contrast, fishes in the entire Internet for each individual word. It is very important therefore, to specify the search words as precisely as possible.

You will see that many search engines are advertised with the number of listed entries – this may be several million in some cases. But do not be deceived by this. A well-run, editorially managed web catalogue results in significantly fewer search hits than a search engine (also termed **robot-generated index**). However, the listed entries are mostly much closer to the desired subject.

The web catalogue is well suited to searching for a particular topic or subject area. It is less appropriate for clarifying a particular problem or subject. Well-known catalogues include 'Yahoo!', 'Lycos' and 'UK Plus'.

111

You cross your desired topic and delve deeper into the hierarchy of the web catalogue. The more precisely you limit your topic at this point, the more exact will be the results delivered by the web catalogue. Web

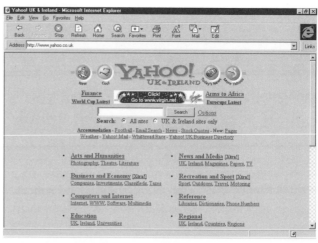

catalogues of this type often offer the possibility of links. This means that you can link two or more search terms together. To find information on the history of the Internet, for example, you have to connect the terms Internet and History together using an **AND operation**, though not all web catalogues offer this possibility. You can find this out from the web catalogue concerned.

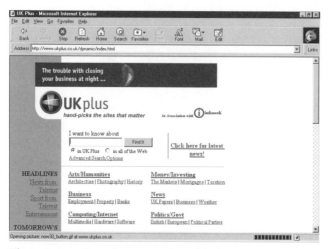

A search engine consists essentially of three components. The first part is the information collector, 'robot', 'spider' or 'crawler' – which moves through the Internet and surfs pages automatically.

The accumulated mountain of information is sent to the management system – the 'index'. The second component is **indexing software**, which structures the data provided and makes it transparent.

The third and final component evaluates the search requests and sends the request to the data computer, to present the search results from there to the user.

A few tips for searching

You will probably know the expression, 'all roads lead to Rome', and experienced Internet users can testify to this. When searching for a particular piece of information on the Internet, you must adjust your **way of thinking** slightly. Get used to thinking on the computer's level. The logical connection of search terms comes to the fore; the material content of the desired topic should drop into the background. In other words, you should think of special generic terms for the topic, using the expanded search options of search engines for this purpose. It would be nonsensical, for example, to search for a term such as 'Internet' or 'salt'. The hits list would run to hundreds of thousands. Instead, you should set or define a field that is related to the desired topic area. If, for example, you want to know about salt extraction from the Red Sea, you should exclude aspects such as 'chemistry', 'minerals' and 'mining' from your search list, since this would contribute too little to the topic. Instead combine the search terms 'salt', 'food', 'Red Sea' or similar relevant terms.

On the Internet, there are various WWW servers which keep contact with all major search engines. You enter a search term just once at this point and this is sent to all corresponding search engines and catalogues, so that you no longer have to comb manually through all the search engines and catalogues. Colleges and universities, in particular, have set up these types of services.

Be careful, however, with your **AND connections**:. Too many connections can result in no hits, since the database of the search engine is not large enough. It is more sensible to limit and vary search terms gradually.

113

Generally, you can connect words together using '+' or 'AND'. This means that the terms connected by these **operators** must occur in the final document. A minus sign '-' or 'NOT' excludes a subsequent word. This word will not be contained in the result. Several words can be connected with inverted commas into one phrase. These are then treated as a single term, e.g. 'Tony Blair' or 'Italian pop music'.

Syntax differs, however, from one search engine to another but most search engine operators instruct you in this.

Using the Lycos search engine

After all that theory, it is time to put what we have learned into practice. You will therefore use a web catalogue for your purposes.

Imagine, for a moment, that you are looking on the Internet for photographs of the current NASA mission to Mars.

Start your Internet Explorer and connect to the Internet. Next select the web site of our web catalogue – *http://www.lycos.com* and enter the search terms 'NASA' and 'Mars'.

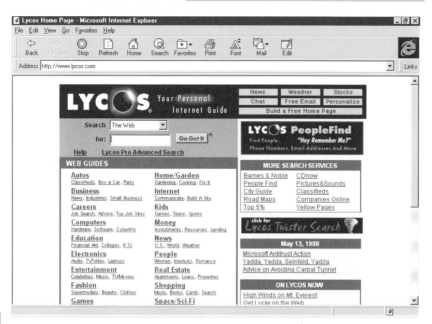

The Lycos web catalogue with
the URL *http://www.lycos.com*.

Enter the terms *NASA* and *Mars*
in the field labelled *for:*, and then
click the *Go Get It* button.

After a short time, a new page will appear containing links to all the
pages Lycos could find containing the two words we searched for.
You can jump to one of these pages by clicking its link on this page.

115

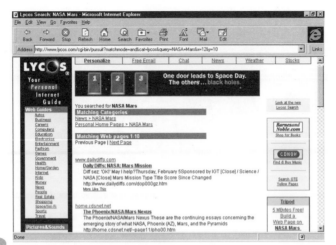

3 A single page of results appears. If you scroll to the bottom of the page, you'll find links to pages containing yet more results for this search.

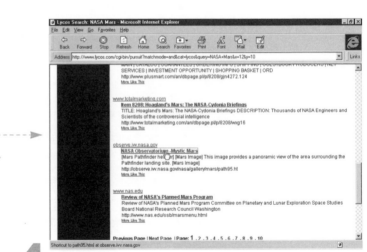

4 Read the short descriptions of each result and click on one that sounds as if it will contain images.

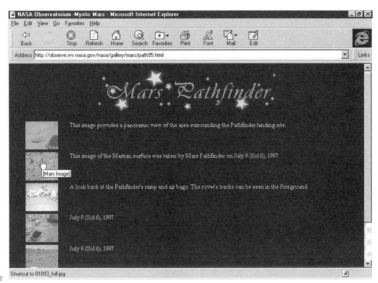

5 Here is the desired page on the Internet, with the corresponding information.

Using the catalogue

Until now, you have entered a **search term** directly to obtain the desired document. Next we'll use a web directory to search for similar types of site by selecting categories. For this we'll use the popular directory, UK Plus.

Start Internet Explorer and type the address *http://www.ukplus.co.uk* into the address field.

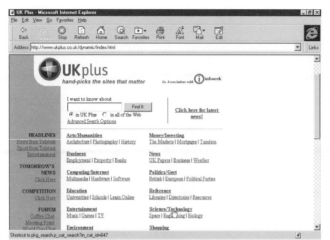

1 Click on the *Science/Technology* hyperlink.

2 Click on the *Space/Astonomy* hyperlink in the resulting page.

3 Click on the *Astronomy* catalogue entry.

4 This leads to a page containing links to 10 astronomy-related sites.

If you can't find the type of site you're looking for among the 10 links on this page, just scroll to the bottom of the window. Here you can see that UK Plus currently has 57 astronomy-related links in its catalogue, and you can get descriptions of the next 10 and links to them by clicking the *Next Page* link. Do this on each page until you've found the type of astronomical site you want (or exhausted all 57 possibilities!).

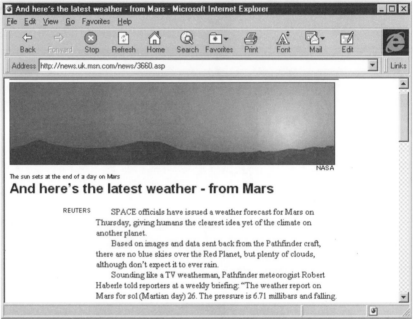

And here's the latest weather - from Mars

REUTERS   SPACE officials have issued a weather forecast for Mars on Thursday, giving humans the clearest idea yet of the climate on another planet.

Based on images and data sent back from the Pathfinder craft, there are no blue skies over the Red Planet, but plenty of clouds, although don't expect it to ever rain.

Sounding like a TV weatherman, Pathfinder meteorologist Robert Haberle told reporters at a weekly briefing: "The weather report on Mars for sol (Martian day) 26. The pressure is 6.71 millibars and falling.

5 Here it is, more Mars information.

Through the depths of the Internet with Gopher

A somewhat older, but interesting option for finding documents on the Internet is the Internet Gopher service. Gopher first emerged in 1991 at the **University of Minnesota**. The objective was to develop a tool that enabled simple access to various resources on the Internet.

In contrast to the Internet services you have met so far, you do not need any details for Gopher, such as file names or host addresses, to search for particular pieces of information. To launch a successful search, only the Internet address of the desired Gopher server is needed.

The operation of a **Gopher server** is simple in principle: Gopher makes a menu system available to the Internet surfer, under which the various Internet resources can be summarised. Access to these resources is direct from the Gopher interface.

Most Gopher servers on the Internet are connected to one another, so that the input of a search term is passed from one Gopher server to the next, until all the possibilities have been exhausted. During your Gopher expeditions through the Internet, you will probably run into the term **Gopher Space** from time to time. Under this heading are all of the documents on the Internet managed by Gopher servers.

The following table includes the Internet addresses of some important Gopher servers on the Internet, which facilitate your path into Gopher Space.

Gopher	Location
MICROS.HENSA.AC.UK	United Kingdom
CAT.OHIOLINK.EDU	USA
FINFO.TU.GRAZ.AC.AT	Austria
GOPHER.TORUN.EDU.PL	Poland
GOPHER.SUNSET.SE	Sweden

There are countless Gopher servers on the Internet. Once you have registered with a server, you can switch from Gopher server to Gopher server.

To participate in Gopher, you need a Gopher client. Masses of such programs can be found on the Internet and cost only a few pounds.

For our purposes, the program **WSGopher** is the most suitable. It has won great acclaim in the Internet community in recent years and is in widespread use. Particularly appealing is that in the software, there is an extensive range of interesting addresses in Gopher Space, which are organised according to subject and can be activated by the click of a mouse.

Downloading WSGopher from the Internet

In this case also, you can find all you need from the Internet. Make a connection to the Internet and enter the following URL on Internet Explorer: *http://www.shareware.com.*

On this web site, you will find a variety of useful and helpful Internet programs for which you do not have to pay a penny – they are available to download free of charge.

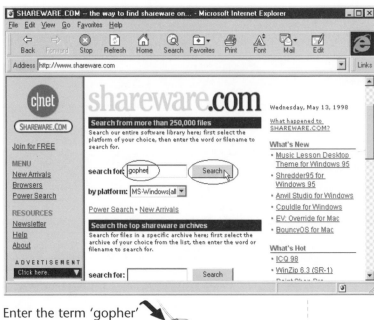

1 Enter the term 'gopher' in the input field and confirm using the *Search* button.

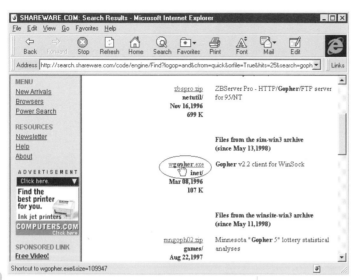

2 Click on the entry WGOPHER.EXE.

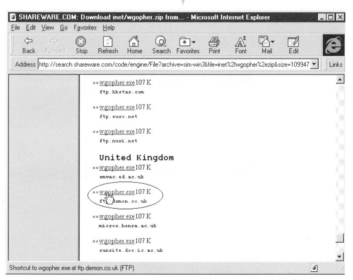

3 Scroll down the page to find a download site in the UK and click one of these links to download the program.

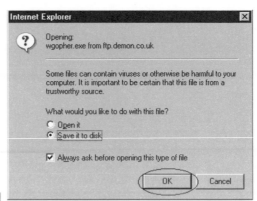

4 Activate the checkbox *Save it to Disk* and next press the *OK* button.

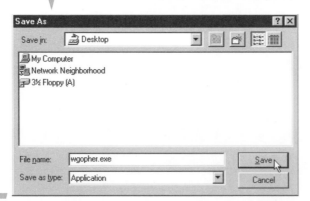

5 Choose a directory on your hard disk, into which the corresponding file is to be saved. Next, click on *OK*.

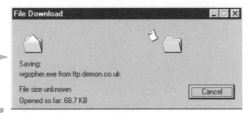

6 A dialog will keep you informed about the download progress.

After the download process is complete, there is a file in your destination directory on the hard disk, which must first be **decompressed** in the next step. This file is a ZIP file, a compressed file that requires a certain program to decompress its contents. The file from our example is a type that decompresses itself automatically when executed. Open either the MS-DOS Prompt or the Windows 95 Explorer to start the file.

Along with decompressing the file, we will also create a shortcut to WSGopher on the Windows desktop for easy access. Since the Gopher program does not perform an installation, you must do this manually.

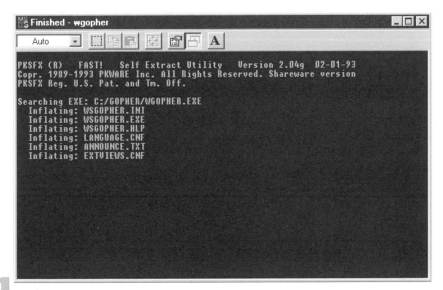

1 Double-click WGOPHER.EXE to extract its contents.

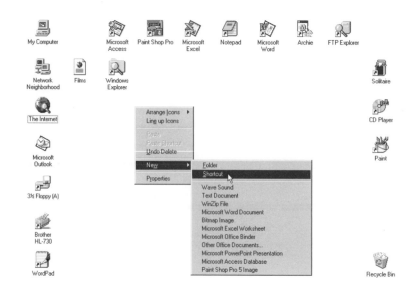

2 Click the right-hand mouse button on the Windows 95 desktop, move to NEW and click on SHORTCUT.

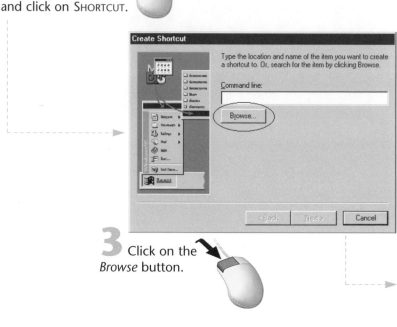

3 Click on the *Browse* button.

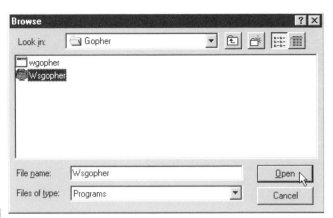

Choose the directory in which the Gopher file is located.

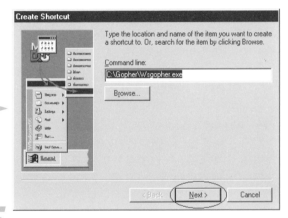

The program has been found, so finally click on the *Next>* button.

127

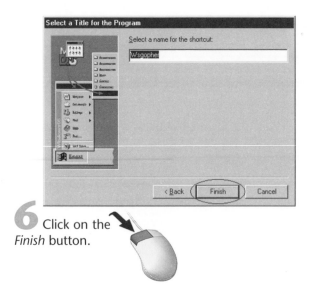

 Click on the
Finish button.

Once this procedure has been completed, you will notice that a new icon has appeared on your Windows 95 Desktop. In the next step, you will start and operate the program.

First steps into Gopher Space

After you have installed the Gopher program on your PC, it is time to venture into Gopher Space. Make a connection to your Internet service provider, and once a connection has been successfully made, click on the Gopher icon on your Windows 95 Desktop. As soon as the application opens, you will define a Gopher Home Server. This is the Gopher server that is selected first when the program is started.

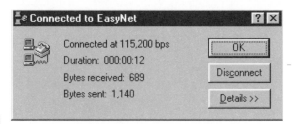

Make a connection to your
Internet service provider.

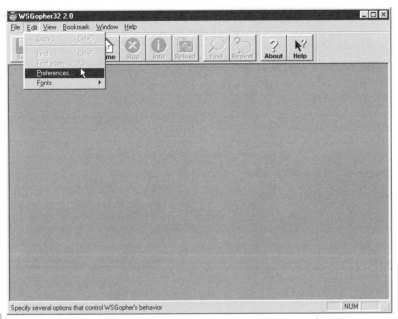

2 Click on the menu item
PREFERENCES from the Edit menu and
then select the HOME GOPHER tab.

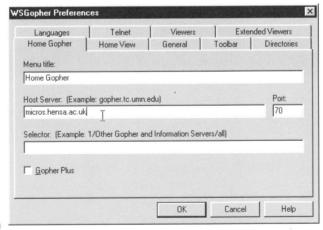

3 Enter the name of the Gopher server in the
Host Server input field, which will always be
called upon first when the program is started.

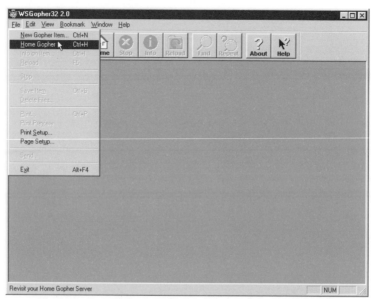

4 Start the HOME GOPHER menu item from the FILE menu to connect to the corresponding Gopher server.

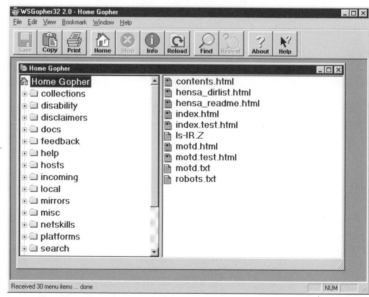

5 After a few moments, you are connected to the home Gopher server.

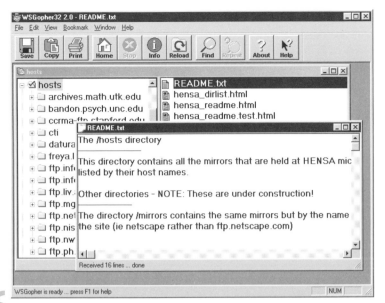

6 At the click of a mouse, you have the
opportunity to move through the various menus.

Bookmarks simplify life for you

On expeditions through Gopher Space, you will quickly meet a
range of **Gopher server addresses** on the widest range of topics. To
keep note of the most interesting and important Gopher servers, you
have the possibility of recording these in your personal address
collection.

WSGopher already has in its default settings a comprehensive
bookmark collection. As you can see, the dialog box consists of two
separate sections. In the left-hand window, you can see the
individual categories. Click with the mouse on the corresponding
entry in the left-hand field, so that the entries stored in the
categories appear in the right-hand field. Every sub-item in the right-
hand section of the window represents its own Gopher server, which
you can reach via the Internet.

To reach other Gopher servers, you have to click on an entry in the
left-hand section of the bookmark list.

To examine the existing **bookmark list**, follow the steps given below:

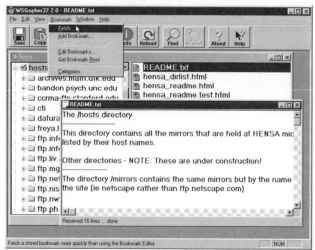

1 Click on the FETCH item from the BOOKMARK menu.

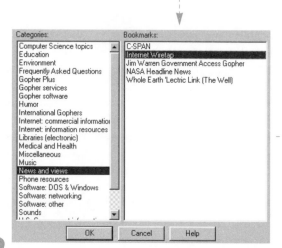

2 You see a list of the bookmark entries already integrated. Click on *News and Views* and on the bookmark entry *Internet Wiretap*.

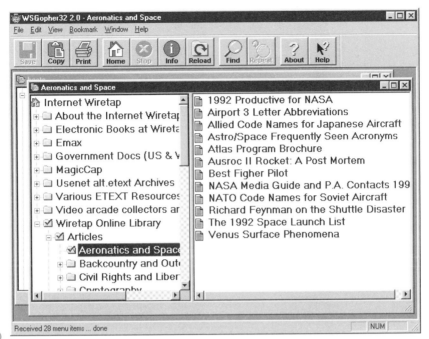

3 You can move effortlessly from Gopher server to Gopher server using the bookmark list.

Producing your own Gopher bookmark entries

You can, of course, also produce your own bookmark entries. This is particularly useful, if you are in Gopher Space and collect interesting addresses and first save these in a general directory in the bookmark list. At a later date, you can then arrange all the entries.

To produce a new bookmark category, you have to select the CATEGORIES entry from the BOOKMARK menu.

In the following dialog box, you find all existing categories. In order to produce a new category, you enter the desired name in the input field and then press the *Create* button.

133

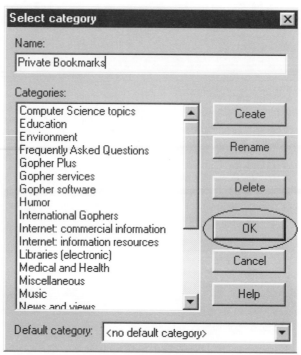

1 Enter *Private Bookmarks* – the name of the new bookmark category – and click on *OK*.

So that all the addresses found in this category, during the Gopher sessions are saved, you must enter your new bookmark group in the DEFAULT CATEGORY pull-down menu.

Selecting the EDIT BOOKMARK item from the BOOKMARK menu gives you the possibility of editing your entries and moving them to another category.

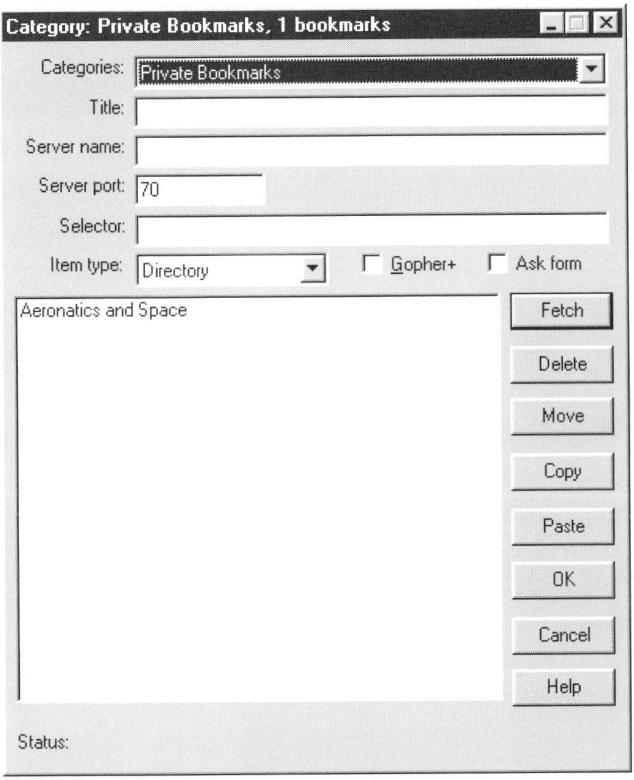

Category: Private Bookmarks, 1 bookmarks _ □ ✕

Categories:	Private Bookmarks ▼
Title:	
Server name:	
Server port:	70
Selector:	
Item type:	Directory ▼ ☐ Gopher+ ☐ Ask form

Aeronatics and Space

- Fetch
- Delete
- Move
- Copy
- Paste
- OK
- Cancel
- Help

Status:

2 At this point, you have the option of editing your bookmarks and moving them to other categories.

135

5

**Music from the
World Wide Web**

What's in
this chapter?

The World Wide Web is an interactive source
of multimedia data. You can hear music, news
or speech over the WWW and play it through
the speakers connected to your computer's
sound card. In this chapter you will learn how
to make use of this fascinating resource. You
will need to install a 'Real Audio player' to
enable Internet Explorer to play
speech and music files live
from the Internet.

You already know:

You are going to learn:

Rock, pop and classical music from the Web

We make use of all our senses to take in the world around us. Our eyes and ears play a primary role – they are our most powerful sense organs. Since the invention of the World Wide Web the Internet, the modern communications medium, has plenty to offer our eyes and ears.

Sounds, voices and music have meanwhile become an everyday part of the Internet – from radio stations, interviews on the Web and music to sound files and full-length live concerts, all of which are transmitted over the WWW.

WHAT'S THIS?

Digitising: Analog data, such as speech or video, has to be converted into ones and zeros before the computer can understand it. This conversion process is called digitising.

On the Internet there is a vast number of different audio types. They all have one thing in common: they have been digitised so that they can be transmitted over the Internet. You have probably already come across files with the ending .WAV or .AU. These are files that contain audio data such as music or speech. To play these on your computer, you need a special program known as an audio player. If you have Windows 95, you already have one of these.

As a rule, audio files are very large. The first thing you must do is download them to your computer from the Internet. You can only listen to them after they have been transferred in full onto your hard disk. Obviously this is very time-consuming. It is not at all unusual to have to wait more than 20 minutes for a file that may contain only one minute of music!

No matter. A far more efficient and modern application of audio on the Internet is available called 'streaming audio'. This processes audio data intelligently. You no longer have to wait until the whole of the file has been transferred before you can play it. You can already hear the sounds during the download.

For this you need a special program that is automatically started up by your Web browser as soon as you select a hyperlink containg a corresponding sound file. The program that you need is called 'RealAudio player' and is, of course, freely available on the Internet.

How does RealAudio work?

When you click on a link to a RealAudio file within a Web site, this link does not go directly to the file concerned. Instead, your browser sends a request to the WWW server and this sends back a **RealAudio metafile**. This is a small text file that contains the address (URL) of the file. It also contains instructions which the Web browser can use to start the RealAudio player automatically if an error should occur. When the RealAudio player has been started, it tries to reach the URL in the metafile. This URL, however, is not located on the corresponding Web server but on a special RealAudio server which has the specialist task of supplying RealAudio files to users.

A **RealAudio** server holds music or speech files which you can call up over the Internet.

The RealAudio server uses the RealAudio player to determine the speed of your Internet connection. If it finds that this is slow it sends a file with low sound quality. If you have a fast connection the sound quality is high and more files have to be transmitted to achieve this.

Buffer is the name given to a special memory area in which your computer stores data temporarily.

The RealAudio file that is to be transferred has been compressed so that it can be sent more quickly. At the receiving end, all incoming data packets are placed in a special storage area known as a buffer. As soon as this buffer is full, all the files are sent to the RealAudio player.

With RealAudio you can wind forward or back within an audio file. If you do this with a file of this type, the player informs the server of the new position and the RealAudio server sends the file from the new starting point.

139

Installing a RealAudio player

To experience the advantages of sound transmission from the Internet you must install a RealAudio player. Like everything else, you can get this from the Internet.

Start up Internet Explorer and input the URL *http://www.real.com/ products/player/index.html*

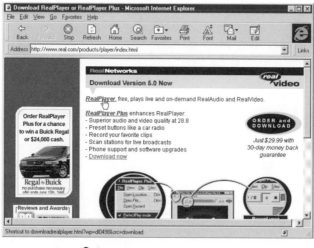

1 Click on the RealPlayer hyperlink.

Once the new window has opened you must first type into it the operating system that you have installed, the processor that is in your computer and your email address. When all this information is complete, you can click on the *Download Now!* button.

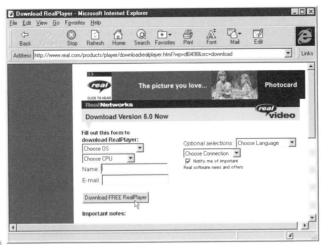

2 Fill in the brief form and click the Download button.

You can now select a WWW server from which to load the RealAudio player onto your computer. In our case you must find out which server is the fastest. If you happen to be surfing during the night, I would recommend a WWW server in Europe as at that time the Internet is not necessarily as overloaded as during the daytime, for example.

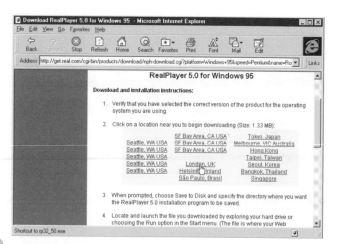

RealPlayer 5.0 for Windows 95

Download and installation instructions:

1. Verify that you have selected the correct version of the product for the operating system you are using.

2. Click on a location near you to begin downloading (Size: 1.33 MB):

	SF Bay Area, CA USA	Tokyo, Japan
Seattle, WA USA	SF Bay Area, CA USA	Melbourne, VIC Australia
Seattle, WA USA	SF Bay Area, CA USA	Hong Kong
Seattle, WA USA		Taipei, Taiwan
Seattle, WA USA	London, UK	Seoul, Korea
Seattle, WA USA	Helsinki, Finland	Bangkok, Thailand
	São Paulo, Brasil	Singapore

3. When prompted, choose Save to Disk and specify the directory where you want the RealPlayer 5.0 installation program to be saved.

4. Locate and launch the file you downloaded by exploring your hard drive or choosing the Run option in the Start menu. (The file is where your Web

Shortcut to rp32_50.exe

3 Click on one of the hyperlinks.

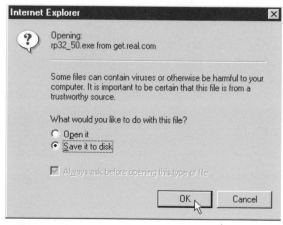

Internet Explorer

Opening:
rp32_50.exe from get.real.com

Some files can contain viruses or otherwise be harmful to your computer. It is important to be certain that this file is from a trustworthy source.

What would you like to do with this file?

○ Open it
● Save it to disk

☑ Always ask before opening this type of file

OK   Cancel

4 Click OK and then choose a folder in which the downloaded file should be placed.

5 The download begins.

When the download is completed, start up the file from the relevant folder on your hard disk.

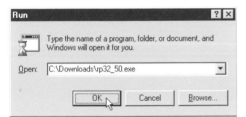

1 In the Run dialog, click on the Browse button to find the correct file, or enter the file name directly in the field Open. Click on OK to start the install program.

2 Click on the Next button.

143

3 Accept the conditions of the licence agreement by clicking on the *Accept>* button.

In the next stage of installation you have to input your name, company name if appropriate and email address. This is so that you can be informed by electronic mail of any new developments or updates.

A window then appears in which you are asked to input the speed of your modem.

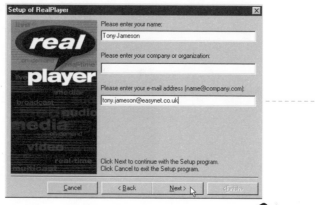

4 Enter your name, your company (if apporpriate) and your email address, then complete the input by clicking on the *Next>* button.

5 From the list, select the speed at which you are moving around the Internet. Click on *Next>*.

In the next step the installation routine asks you for a destination folder into which the program will be installed. At this point, if you want to select a different folder from the one suggested, you can use the Browse button to move through the tree of folders on your hard disk.

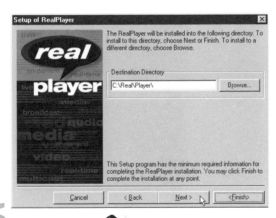

6 Click on *Next>*, to accept the folder that is suggested.

145

Now the installation routine checks to see if a Web browser is already installed on your computer. In our example, it finds Microsoft Internet Explorer and marks this for use with the RealAudio player. If you have other Web browsers installed, for example Netscape Navigator, you must activate all the browsers that will use the RealAudio player.

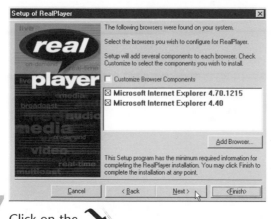

7 Click on the *Next>* button.

8 Click on the *Finish* button to complete the installation.

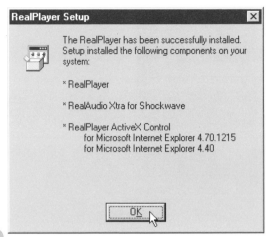

RealPlayer Setup

The RealPlayer has been successfully installed. Setup installed the following components on your system:

* RealPlayer

* RealAudio Xtra for Shockwave

* RealPlayer ActiveX Control
   for Microsoft Internet Explorer 4.70.1215
   for Microsoft Internet Explorer 4.40

OK

9 The RealAudio player has been successfully installed. Click on *OK*.

Once you have clicked on the OK button, Internet Explorer starts with a URL which takes you a little deeper into the subject of RealAudio.

Using the RealAudio player

After you have installed the RealAudio player it is time to take the plunge and call up some sounds from the Web browser.

Start up Explorer and enter the following URL: *http://www. liveconcerts.com.* This is a site that broadcasts pop concerts in RealAudio. But if you're not lucky enough to arrive while a concert is going on, you can still listen to individual songs.

1 Click on one
of the links below
Listening Parties.

2 Click the *RealAudio*
icon to hear short clips of
the chosen songs.

Once you have activated a piece of music, the RealAudio player is called up by Internet Explorer and the sound file is played during the download procedure.

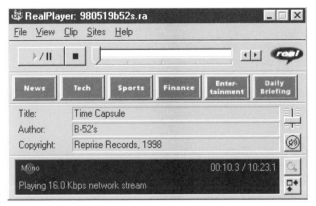

3 The RealAudio player automatically starts playing the piece of music you have chosen.

The Internet has a large number of Web pages with RealAudio files. Use a Web catalogue to sift through the Internet using the relevant search terms.

The RealAudio player in detail

Now that you have experienced for yourself how a RealAudio player works, it is time to have a closer look at the player and its functions.

After you have installed the RealAudio player you will find an icon on your Windows 95 desktop labelled RealPlayer and a folder of files on the hard disk with the name *C:\Real*.

Double-click on the *RealPlayer* icon on your screen. When the program starts you will go into a registration screen that asks you if you would like to register your copy.

If you do this you will get the latest program information at regular intervals, or, for example, updates to later versions. But in our case, we will skip this stage.

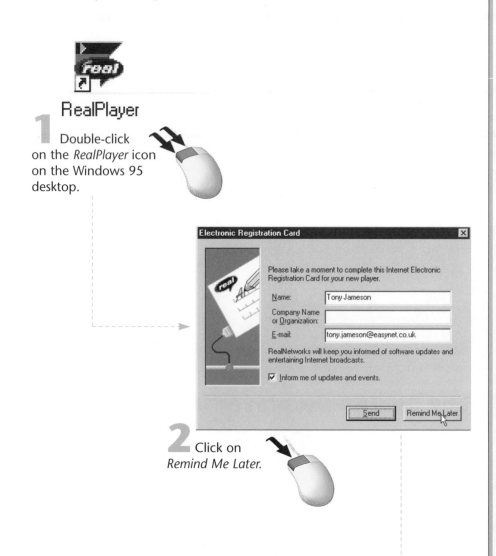

RealPlayer

1 Double-click on the *RealPlayer* icon on the Windows 95 desktop.

2 Click on *Remind Me Later.*

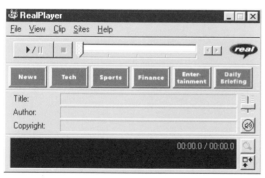

3 RealPlayer is now open.

As you can see, the program starts with a window displaying various functions which are not immediately obvious at first glance.

The central component of the RealAudio player consists of six large buttons with a blue background, which bring you to the 'RealAudio' company's Web server. Behind these buttons are hidden lots of pages that all contain RealAudio files. You are going to follow up a hyperlink – **News**. Click on the News button and see what happens.

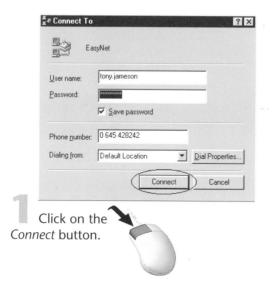

1 Click on the *Connect* button.

151

2 The RealAudio player has taken you to the corresponding Web page. Click on one of the icons beneath *Try A Sample.*

3 The RealAudio player loads the data from the RealAudio server.

When the loading process is complete, the RealAudio player automatically starts playing the current news item – in this case, the student protest in Jakarta.

4 The news is automatically played.

In the fields *Title, Author* and *Copyright* you are supplied free of charge with the information about the author, the title and of course the **copyright**.

To the right of these fields there is a slide control that you can use to adjust the volume of the news item.

6

Looking
into movies

What's in
this chapter?

If you thought that you had reached the
limits of what is technically possible when you
transmitted sound from the World Wide Web,
then you were wrong. There is one step
more. This workshop is going to show you
how to capture video
sequences over the
Internet. In this
chapter you will
find out what
additional
program you
need to watch
movies over the
Internet. You
load this
software from
the Internet
and install it
on your computer.
Finally, a look at a movie
preview site will show you how it works!

You already know:

You are going to learn:

Video on the Internet – how does it work?

Video sequences are short animated films which you can call up over the Internet.

Over the course of time the Internet has moved away from being just a text transmission system (email and Telnet) and developed into a graphic medium (the World Wide Web). The logical consequence of this is that video sequences or even full-length films or documentaries can be transmitted over the World Wide Web. But that is still in the future. For the moment, short **video sequences** are arriving on the Net just to give us an idea of how it could look in ten years' time.

In the meanitme developers and engineers are working feverishly on the probelm of shortage of bandwidth. Accumulation techniques known as streaming video are already in use. They tackle a basic problem head on. As a rule a video contains a lot of data, video data, waiting to be transmitted over the Internet in the form of a correspondingly large file. Sometimes this could take hours.

Streaming video tackles the problem in two ways: first, the video data is very heavily compressed so that less information is waiting to be transmitted over the Net; second, the videos can be played while they are being transmitted.

How exactly does this work? Well, streaming video means that you can watch videos live on the Internet. You do not have to wait until the whole video file is on your hard disk. An example of streaming video is **VDOLive**, which we'll download and use in this chapter. Before it is transmitted the video is compressed and encoded. This process takes place by means of a special algorithm (a mathematical formula) which makes the video file considerably smaller.

When you are exploring the Web, if you come across a page and click on a hyperlink that leads to a video, a message is sent to a server to request the video.

The corresponding file is then sent over the Internet. The video data is sent to a special storage area in your computer, also known as a **buffer**, which is usually between 5 and 30 kilobytes in size. The server that is sending the data calculates the transmission speed of the connection from the rate at which the buffer fills up. At high speed, more video data is transmitted and the video is more lifelike. Low transmission speeds mean poorer video quality.

As the buffer is being continuously filled, which only takes a few seconds, your computer starts up a **video player**, which begins to play the video. While you are watching the video, more data is transmitted to the buffer and at the same time video data is sent from the buffer to the video player. After all the video data has been transmitted, the display terminates. The video file is not stored on your computer but is deleted as soon as it has been played.

Installing a video player

Well, now you know everything you need to know about the technical aspect of video transmission over the Internet. Once again, it is time now, once again, to put the theory into practice. Begin by downloading the video player from the Internet. To do this, go onto the Internet and visit the VDOLive site: *http://www.clubvdo.net/clubvdo.*

Once you have successfully loaded the page that you need, you can start downloading the VDO video player.

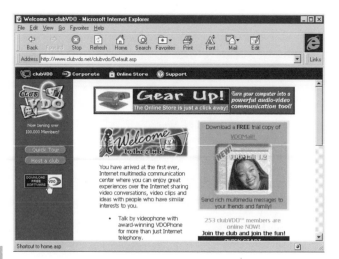

1 Click on the *Download* button.

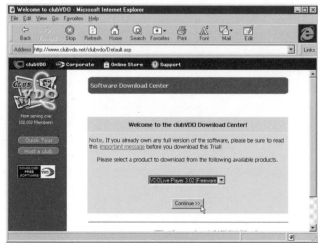

2 Choose the *VDOLive Player* from the drop-down list and follow the remaining simple steps, clicking the *Continue* button to move on after each.

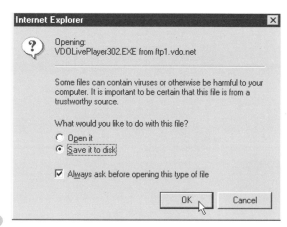

3 The file is now loaded from the Internet. Activate the radio button *Save* and then click on the *OK* button.

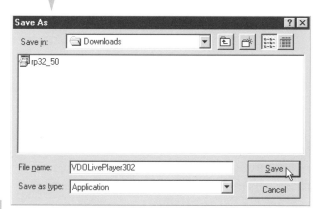

4 Choose a folder on your hard disk for the file to be copied to from the Internet. Then click on *Save*.

Once you have selected a folder on your computer for the file to go into from the Internet, the **download** from the Internet begins. You can now take a break, sit back and relax.

159

5 The download from the Internet begins.

After the file has been transferred from the Internet to your computer it has to be **unpacked**. The name of this file ends in *.exe*, so all you have to do is run the file. The simplest way of doing this is to click on the *Start* button on the Windows 95 desktop and then go to the menu command RUN.

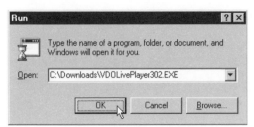

1 Type the path to the file or trawl through your hard disk by clicking on the Browse button. When you have located the file, click on *OK*.

2 Click on the *OK* button.

In the next stage of installation you are required to read the **licence conditions** and, if you agree with them, accept them.

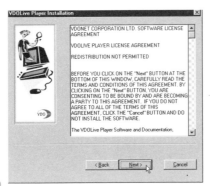

3 Click on the *OK* button.

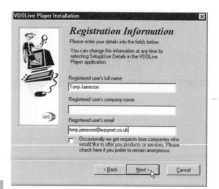

4 Enter your name, company and email address. (You can just type a space in the COMPANY field.) Click on *Next>* to confirm your inputs.

5 Click on the *Next* button to begin the installation.

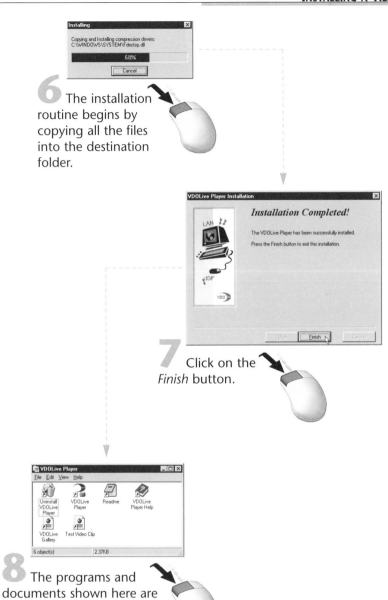

6 The installation routine begins by copying all the files into the destination folder.

7 Click on the *Finish* button.

8 The programs and documents shown here are now on your hard disk.

163

The VDO video player is located in a folder on your hard disk. However, you do not have to worry about this, as the **Web browser**, in our case Microsoft Internet Explorer, has already been given the location of the video player. When you click on a hyperlink which uses the VDO video player, it is automatically **called up**. You don't need to do anything about it. Let's get onto the Internet and test this out at a VDOLive site. Type this address into your browser's address bar: *http://www.famvid.com/nrelease.html*

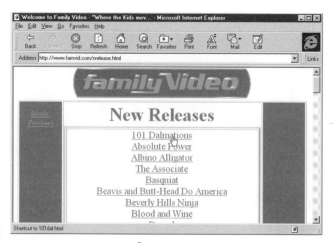

 Choose a movie title from the list on this page.

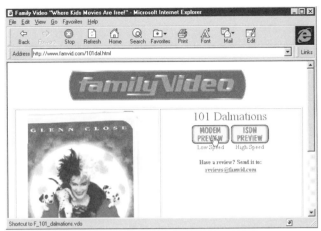

2 Click on the Modem Preview button (or the ISDN Preview if you have an ISDN connection).

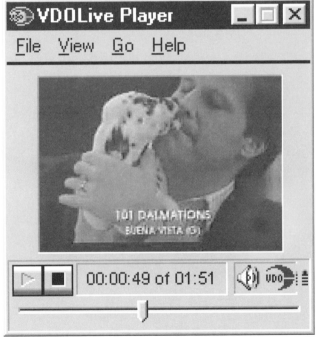

3 Voilà!, live video on the Internet!

165

7

What's in this chapter?

The Internet is a meeting point for all sorts of people from a wide variety of countries all around the globe. In this chapter you will find out about the possibilities for contacting other Internet travellers and hold computer conversations with them. You will load the software that you need from the Internet off the Microsoft Web Server and then install it.

You already know:

You are going to learn:

What is IRC chat?

There is no more direct form of communication on the Internet than taking part in a **chat**. 'Internet Relay Chat', as it is known, allows users to hold conversations over the Internet, not using their voice as they would with Internet telephony, but using the keyboard. The words and sentences that you input using your keyboard are seen by others in **real time** on the screen and vice versa. In this way you can talk to people throughout the world using the keyboard.

Real Time is the name given for virtually immediate transmission of data over the Internet. An example of this is Internet telephony, which normally works without any long pauses.

There are various possible ways of chatting on the Internet and the most popular of these is IRC. Using IRC thousands of people all over the world 'put the world to rights' every day. Each subject has its own **channel**. As soon as you select one of these channels you automatically see everybody else's contributions. But IRC is not just a way of passing the time – it has been positively useful in times of crisis. For example, during the attempted coup in Russia in 1993, eye-witnesses were reporting directly over one of its news channels. It was also used during the great Los Angeles earthquake the same year.

IRC Client: a program which enables you to set up a chat connection with other Internet users.

To take an active part in IRC you need a program known as an **IRC Client**, which receives and transmits all messages to the IRC server on the Internet. Every word that you type is transmitted to a particular IRC server. As a general rule, the IRC servers are not interconnected. 'Your' IRC server sends the message concerned to another server, and this passes it on to all the other users on that server or on the same channel.

After you have set up a connection to an IRC server you must look for a channel (subject area) and a name by which you will be known to others. Every server has a large number of channels and you will

Begin by listening in to the conversations for a while to get a feeling for the subjects and the people present.

notice this immediately the first time you visit. As soon as you enter a channel (also called **chat room** in 'insider circles') you can take an active part in the ongoing conversation. As soon as you have keyed in a message your chat program sends it to the IRC server that you are connected to. The server passes it on to everyone in the chat room.

Installing the Microsoft Chat program

As we have just said, you need a special program to take part in a chat event. You will no longer be surprised to learn that you can get this from the Internet. There is a program on the Microsoft Web server which shows a conversation with people chatting online in the form of a comic strip.

Time to download the program from the Internet. Go onto the Internet and enter the URL for Microsoft's download page *http://www.microsoft.com/msdownload* on your browser.

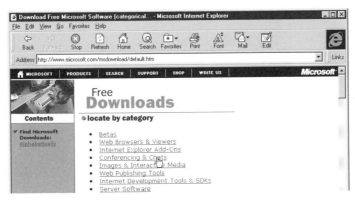

1 Click on the *Conferencing & Chats* hyperlink.

169

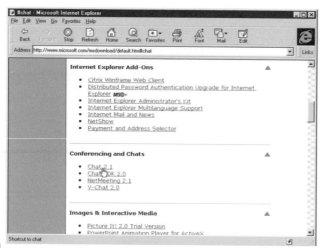

2 Under *Conferencing and Chats* select the hyperlink *Chat 2.1*.

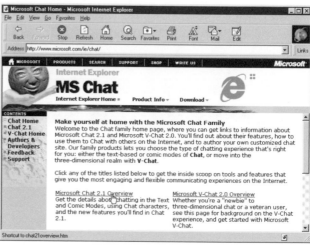

3 Click on the line *Microsoft Chat 2.1 Overview.*

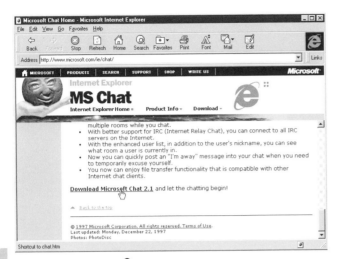

Go to the bottom of the page and click the *Download* link.

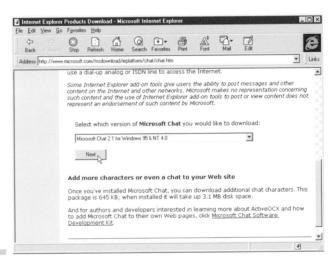

Choose the correct version and click on *Next*.

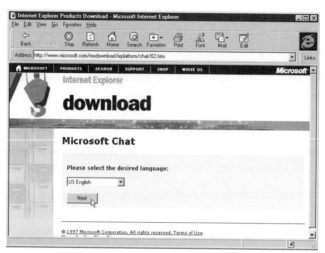

6 Click on
Next again.

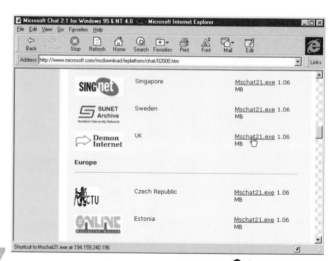

7 Look down the page for a local link to
the file and click it to start the download.

This file will take some time to download, so once again it is time to
take a little break.

Once the loading process is complete you must install the program on the computer. To do this, click on the Start button on your Windows 95 desktop and then activate the menu item RUN. Look for the file in the folder where you stored it.

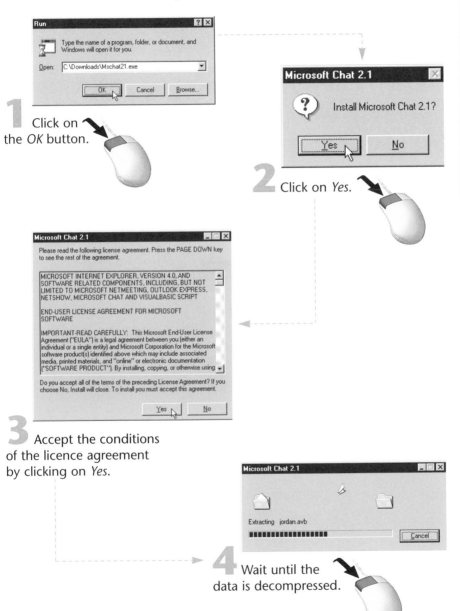

1 Click on the *OK* button.

2 Click on *Yes*.

3 Accept the conditions of the licence agreement by clicking on *Yes*.

4 Wait until the data is decompressed.

173

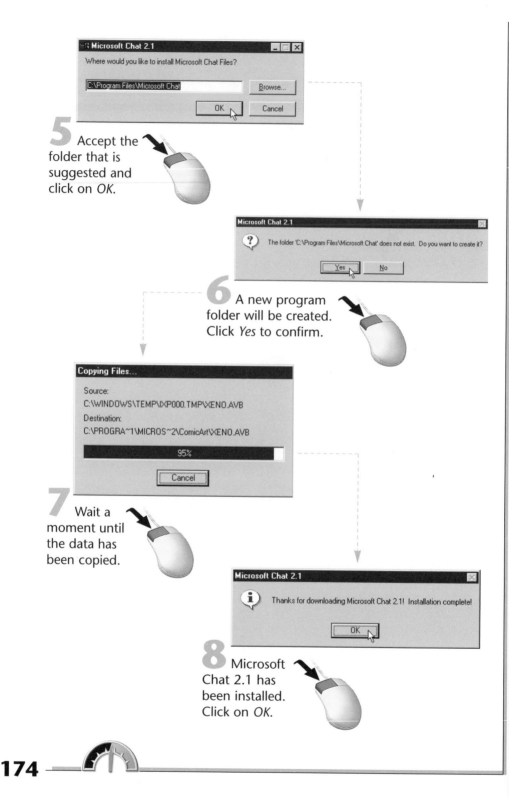

Microsoft Chat 2.1

Where would you like to install Microsoft Chat Files?

C:\Program Files\Microsoft Chat   Browse...

OK   Cancel

5 Accept the folder that is suggested and click on *OK*.

Microsoft Chat 2.1

? The folder 'C:\Program Files\Microsoft Chat' does not exist. Do you want to create it?

Yes   No

6 A new program folder will be created. Click *Yes* to confirm.

Copying Files...

Source:
C:\WINDOWS\TEMP\IXP000.TMP\XENO.AVB

Destination:
C:\PROGRA~1\MICROS~2\ComicArt\XENO.AVB

95%

Cancel

7 Wait a moment until the data has been copied.

Microsoft Chat 2.1

(i) Thanks for downloading Microsoft Chat 2.1! Installation complete!

OK

8 Microsoft Chat 2.1 has been installed. Click on *OK*.

The program is now installed on your computer and ready to run.
The icon for it is in the Windows 95 PROGRAMS menu and you access
this by clicking the *Start* button.

In the menu, click on MICROSOFT CHAT.

A window then appears with the IRC server shown. A **Chat Room** is
already suggested and you are now going to visit this.

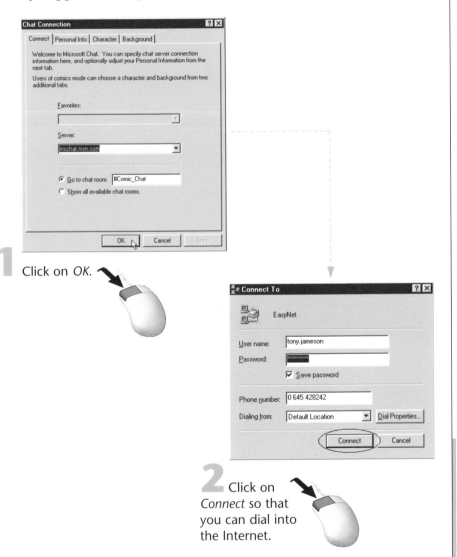

Click on *OK*.

Click on
Connect so that
you can dial into
the Internet.

3 The program is called up.

Now we're on to the serious stuff – you are in a Chat Room. Find out if there is somebody in the room that you can talk to.

Type a comment into the input line at the bottom edge of the window and greet one of the people chatting.

Your greeting will appear in a speech bubble from a cartoon character. As soon as someone reacts to your message a new picture will be added. A comic strip containing all the messages from the channel users will gradually develop.

1 Have a little chat about a subject of your choice.

When you wish to leave the current Chat Room all you have to do is select the menu item LEAVE ROOM from the ROOM menu.

Let's visit a different Chat Room now. First of all get the computer to display a list of all the Chat Rooms available.

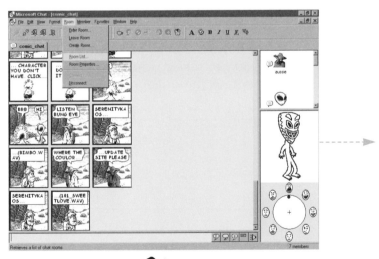

1 Click on the ROOM LIST command in the ROOM menu.

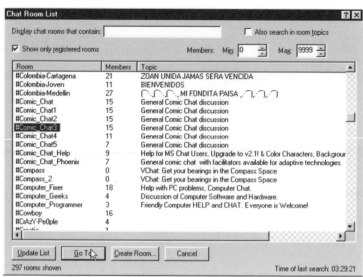

Chat Room List

Display chat rooms that contain: [   ]   ☐ Also search in room topics

☑ Show only registered rooms   Members: Min: 0   Max: 9999

Room	Members	Topic
#Colombia-Cartagena	21	ZOAN UNIDA JAMAS SERA VENCIDA
#Colombia-Joven	11	BIENVENIDOS
#Colombia-Medellin	27	⌒⌒⌒ MI FONDITA PAISA ⌒⌒⌒
#Comic_Chat	15	General Comic Chat discussion
#Comic_Chat1	15	General Comic Chat discussion
#Comic_Chat2	15	General Comic Chat discussion
#Comic_Chat3	15	General Comic Chat discussion
#Comic_Chat4	11	General Comic Chat discussion
#Comic_Chat5	7	General Comic Chat discussion
#Comic_Chat_Help	9	Help for MS Chat Users, Upgrade to v2.1! & Color Characters, Background
#Comic_Chat_Phoenix	7	General comic chat with facilitators available for adaptive technologies.
#Compass	0	VChat: Get your bearings in the Compass Space
#Compass_2	0	VChat: Get your bearings in the Compass Space
#Computer_Fixer	18	Help with PC problems, Computer Chat.
#Computer_Geeks	4	Discussion of Computer Software and Hardware.
#Computer_Programmer	3	Friendly Computer HELP and CHAT. Everyone is Welcome!
#Cowboy	16	
#CrAzY-PeOple	4	
#Croatia	1	

[Update List] [Go To] [Create Room...] [Cancel]

297 rooms shown   Time of last search: 03:29:21

2 Click on a Chat Room
and then on the *Go To*
button.

How to modify the program

Now that you know how to chat it is time to modify the program a
little. At this point you can set up other cartoon characters,
backgrounds and personal information.

Start the Chat program.

1 In the VIEW menu, click on the OPTIONS command.

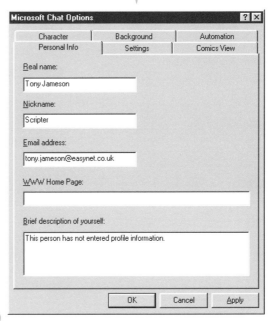

2 Enter your personal details here and confirm by clicking on *OK*.

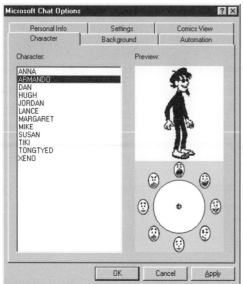

3 Select the cartoon character that you like best and then click on *OK*.

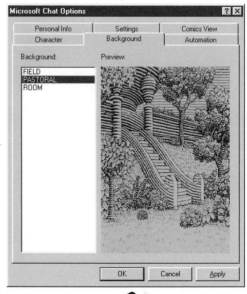

4 Now select a suitable background and click on the *OK* button.

8

Telephoning over the WWW

What's in this chapter?

What was unthinkable only a few years ago has already become reality. It concerns a classic communications tool – the telephone. In this chapter you will learn the technical background to telephoning over the Internet. You will load a program onto your computer from the Microsoft WWW server and with this you will be able to telephone over the Internet. This program can also do a lot more. It can even set up video conferences.

You already know:

You are going to learn:

183

Telephoning over the Web – how does it work?

The Internet has broken away from the standard methods of communication and established new ways for us to talk to each other. You only have to think of email, IRC Chat or Newsgroups (more about these later). Development did not even stop when it came to the classic tool for communication, the telephone. The revolutionary thing about this is that unlike standard telephone calls, you do not have to pay any **telephone charges**. You pay for your local connection to the Internet Service Provider and that's all. Even if you use the Internet to talk to Australia for hours, which would normally cost an arm and a leg, you only pay for the connection to the nearest Internet **access point**.

WHAT'S THIS?

A **sound card** provides your computer with 'a mouth' and 'ears'. This plug-in PC board lets you play and receive music and speech.

To be able to telephone over the Internet you need special software which largely replaces the function of the telephone. You also need a sound card in your computer that has a **microphone input**. The sound is output through the speakers that are connected to the sound card. On the Internet there are lots of programs, some of which have to be paid for and some of which are freely available. In this workshop you will be using the Microsoft program 'NetMeeting'. This is available free of charge from Microsoft's site.

TIP

Don't expect anything wonderful. The sound quality of an Internet telephone call is nothing like as good as that of a normal one. However, steps are being taken to supply good quality sound without the need to connect to ISDN lines.

At present the number of people that you can telephone over the Internet is still limited, but all this is going to change radically very soon.

184

You can get everything you need from the Microsoft WWW server. The way a telephone call over the Internet works is basically as follows: when you are on the Internet and wish to make or receive a telephone call, go to the address book in the program concerned. Here you will find a list of all the other people who use this program. When you have found somebody that you would like to call, the program looks for the **IP address** of the user and passes the call on. If this person is on the Internet his or her telephone will 'ring' at the other end. You can now talk to each other using the sound card and microphone.

As soon as you speak into the microphone the program converts your speech into zeros and ones for, as you know, these are all that the computer understands. In addition all the data is compressed so that it can be sent more quickly over the Internet. The program also determines the speed of the current Internet connection. If you have a very fast connection, the **quality** of the speech transmitted by the program will be very good.

In the next step the spoken word is broken up into small individual packets which are now sent over the Internet. When they arrive at the other end the data packets are decompressed so that the recipient of the call can also understand what you have said. It's that simple!

You can get everything you need from the Microsoft WWW server

Theory is one thing, practice is another. With this in mind you will begin by loading a **telephone program** from the Internet. At this point you decide on Microsoft's NetMeeting.

Dial into the Internet and input the following URL in Internet Explorer: *http://www.microsoft.com/msdownload.*

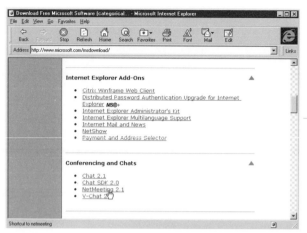

1 Click on
the hyperlink
NetMeeting 2.1.

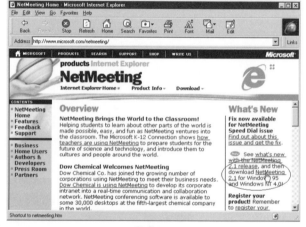

2 Click on the hyperlink
NetMeeting 2.1 to go to
the *Download* page.

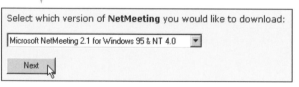

3 Select your operating system –
Windows 95 – and then click on *Next*.

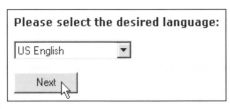

4 Now select the language – in our case English – and click on *Next*.

In the next step you have to select the Internet server from which you would like to load the relevant file. As usual, choose the **closest server** you can find.

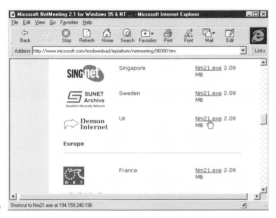

5 Look for an Internet server in the UK.

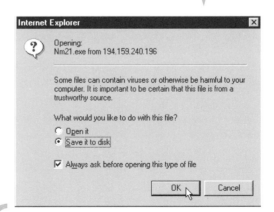

6 Click on the *OK* button to save the file on your hard disk.

187

7 Save the file under the name suggested and select a folder to save it in.

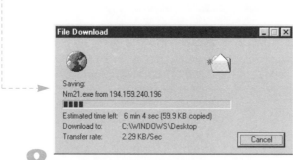

8 Now you will have to wait a while.

Installing the software

Now that the file is on our computer it's time to bring it out of hibernation. The first step is to **unpack** the files that have been transferred and separate them into their individual components. In the Windows Start menu, select the command RUN.

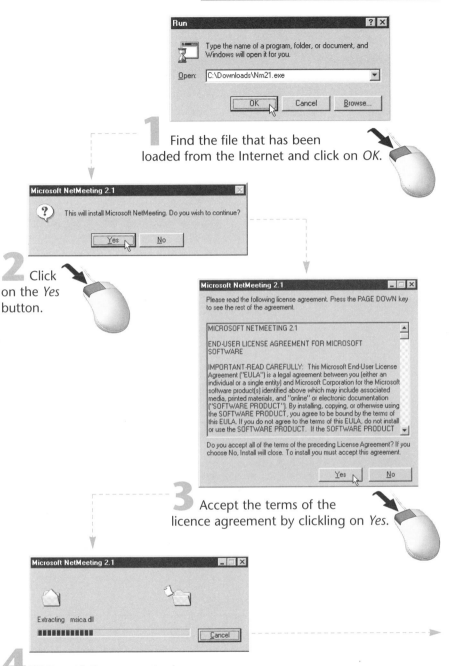

Run ? X

Type the name of a program, folder, or document, and Windows will open it for you.

Open: C:\Downloads\Nm21.exe

OK   Cancel   Browse...

Find the file that has been loaded from the Internet and click on *OK*.

Microsoft NetMeeting 2.1 X

? This will install Microsoft NetMeeting. Do you wish to continue?

Yes   No

2 Click on the *Yes* button.

Microsoft NetMeeting 2.1 _ □ X

Please read the following license agreement. Press the PAGE DOWN key to see the rest of the agreement.

MICROSOFT NETMEETING 2.1

END-USER LICENSE AGREEMENT FOR MICROSOFT SOFTWARE

IMPORTANT-READ CAREFULLY: This Microsoft End-User License Agreement ("EULA") is a legal agreement between you (either an individual or a single entity) and Microsoft Corporation for the Microsoft software product(s) identified above which may include associated media, printed materials, and "online" or electronic documentation ("SOFTWARE PRODUCT"). By installing, copying, or otherwise using the SOFTWARE PRODUCT, you agree to be bound by the terms of this EULA. If you do not agree to the terms of this EULA, do not install or use the SOFTWARE PRODUCT. If the SOFTWARE PRODUCT

Do you accept all of the terms of the preceding License Agreement? If you choose No, Install will close. To install you must accept this agreement.

Yes   No

3 Accept the terms of the licence agreement by clickling on *Yes*.

Microsoft NetMeeting 2.1 _ □ X

Extracting msica.dll

Cancel

4 Wait until the computer has unpacked all the data.

189

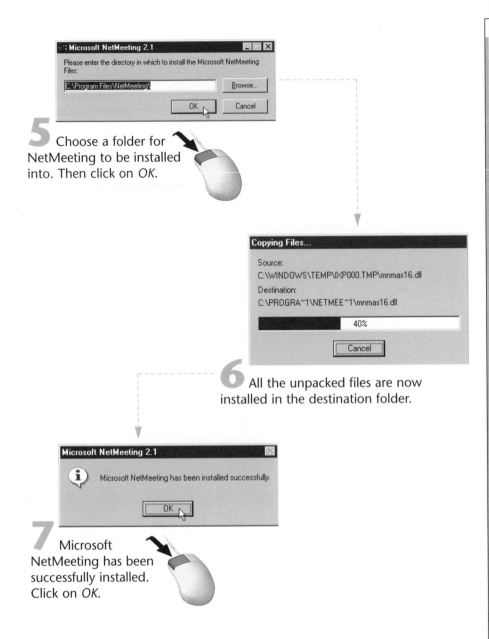

5 Choose a folder for NetMeeting to be installed into. Then click on *OK*.

6 All the unpacked files are now installed in the destination folder.

7 Microsoft NetMeeting has been successfully installed. Click on *OK*.

Now that the software has been successfully installed it can be started up. Click on the Start icon on the Windows 95 desktop and move to PROGRAMS menu. There you will find the icon for the program NETMEETING.

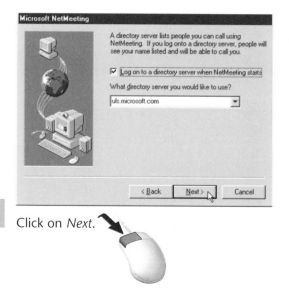

1 Click on *Next.*

The Next step is to create a list of other people who use NetMeeting whom you can contact.

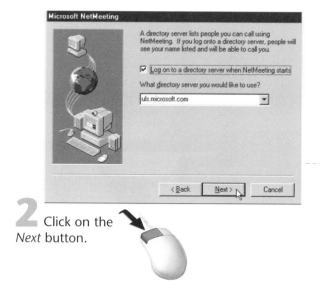

2 Click on the *Next* button.

3 Now enter some personal data so that other people can find you in the directory of participants. Finish by clicking on *Next*.

The installation procedure now asks you which **category of user directory** you would like your details to be stored in.

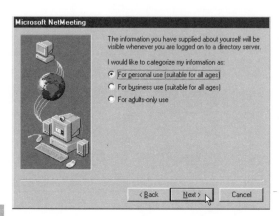

4 Activate the radio button *For personal use (suitable for all ages)*, and then click on the *Next* button.

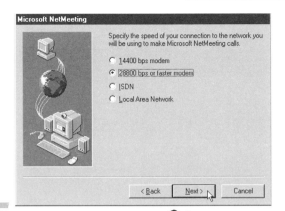

5 Click on the modem speed that you are using. Then click on *Next*.

In the next step of installation, the software identifies the **Microphone** that is connected. To help it do this you have nine seconds in which to say a few sentences loud and clear into the microphone. The aim is to determine how **sensitive** your microphone is.

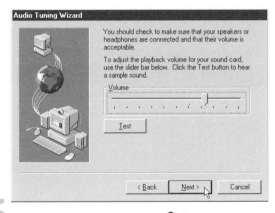

6 Click on the *Start* button, read aloud into the microphone the text that is displayed and finish this step of installation by clicking on *Next>*.

193

7 Click on *Finish*.

How to telephone over the Internet

Now that NetMeeting is installed on your hard disk, there is nothing to stop you telephoning over the Internet to your heart's content.

As soon as you have called up NetMeeting you will notice that the program has a large number of **functions**. You can send emails, have a chat and even create video connections, similar to a videophone. We are only going to look at the telephoning function here. The rest is beyond the scope of this book.

Click on the *Start* button on the Windows 95 desktop and go to the
PROGRAMS menu. In Programs you find an entry entitled NETMEETING.

Select this. Immediately a connection is made to your
Internet Service Provider.

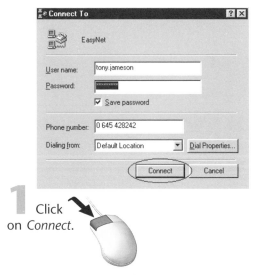

1 Click
on *Connect*.

Once the Internet connection is made, NetMeeting starts up and
immediately connects with a Microsoft Internet Server which has a
directory of all **NetMeeting Users**. This directory is now loaded onto
your computer.

As you can see, a lot of the entries have a little red star above the
computer icon. This means that these people are currently active on
the Internet and you could create a **telephone connection** to them.

195

In the top third of the window you will see two slide controls. The left-hand one controls the sensitivity of the microphone. The right-hand one is for you to adjust the **volume** of your speakers.

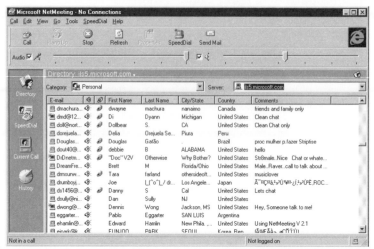

2 Double-click on an entry.

If the person that you would like to telephone is active on the Microsoft Server and is prepared to accept your call there is nothing to prevent a telephone connection.

If you know the email address of the person you want to talk to and they are still online then you can also input the email address directly. To do this, click on the *Call* icon.

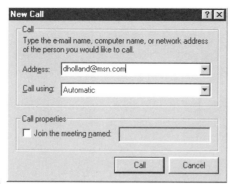

3 Type the mail address of your contact in the *Address* field and then click on the *Call* button.

197

9

Loading a file from the Internet

What's in this chapter?

One of the things you have seen very often in the earlier chapters of this book is how to get a file from the Internet and put it on the computer. However, there is another way of doing this: FTP. In this chapter you will get to know this Internet service a bit better. You are going to download an FTP program from the Internet and install it on your computer. Then you will use FTP to download a file from the Internet. This chapter also shows you something about handling compressed files and how to unpack them.

You already know:

You are going to learn:

199

How does FTP work?

Downloading a file is a standard Internet activity. It enables a great variety of files, such as programs, graphics, sounds, text files and much, much more, to be fetched from the Internet.

Like all the other Internet services that you have already learnt about, you also need a special program for the FTP service. This is called an **FTP Client**. This program makes contact with an FTP server on which another small program runs, making data transfer possible.

To connect to an FTP server in order to download the files, you need an access number or a user name and a password that identifies you as someone authorised to receive data from the server concerned. This procedure has gradually become less and less common over recent years. Most FTP servers now allow anybody access even though a user name and password still have to be entered. With this kind of FTP server, you enter the user name **anonymous** and your email address as the password in order to gain access to the files. These FTP servers are therefore known as 'anonymous FTP servers'.

FTP is extremely easy to use. As soon as you have connected to an **FTP server** you can look at all the available folders and the files that are in them. If you discover something interesting, all you have to do is click on the file concerned and define a destination folder on your computer.

But how does FTP work exactly? To begin an FTP session you must start up the FTP program on your computer and get connected to the Internet. A special program called **FTP Daemon** runs on the FTP server and controls all transfer activities. As soon as you use your FTP program to make contact with an FTP server, the FTP Daemon asks for the user name and a password.

Once you are registered, a **Command Connection** is set up between your computer and the FTP server. The purpose of this connection is to send commands from your computer to the FTP server and receive messages and information.

If you want to change to a different directory in the FTP server, your FTP program sends the appropriate instruction to the FTP Daemon. This changes the directory and uses the command connection to send you a list of the files that are in the new directory.

If you have found a file that you would like to load, a second connection, known as the **Data Connection**, is created. This enables files to be loaded from the FTP server onto your computer. After the data transfer is complete this connection will be automatically closed down. However, the command connection remains active so that after a download you can carry on looking at the contents of the folder on an FTP server.

We need to get an FTP program

There are vast numbers of FTP programs on the Internet and in general they are available free of charge. To get one you visit the WWW server of a company that offers an FTP program for free use.

Set up a connection with the Internet and input the following URL: *http://www.ftpx.com.*

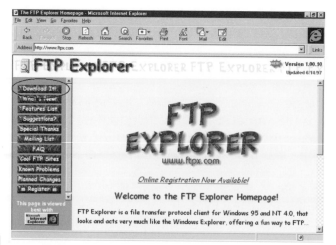

1 Click on the hyperlink *Download It!* in the left-hand section of the Web site.

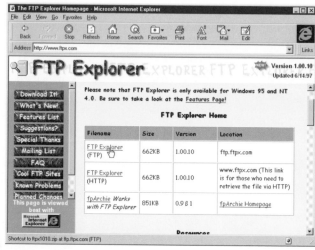

2 Click on the hyperlink *FTP-Explorer* to load the file from the Internet.

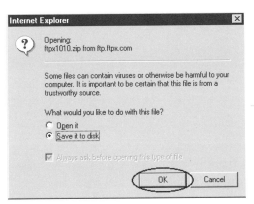

3 Click on the
OK button to
store the file.

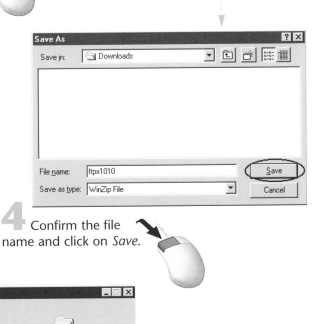

4 Confirm the file
name and click on *Save.*

5 Wait a minute or two until the
file is on your computer.

How to unpack a file

Once the file is on your computer it has to be unpacked. In the previous chapters this always took place automatically. This time the situation is rather different. You are dealing with a file which has the extension *.zip*.

Transferring a file from the Internet takes time – sometimes a lot of time. The bigger the file, the longer the download process lasts. To keep telephone charges as low as possible for Internet users, resourceful minds have invented.ways of **shrinking** files. Once a file has been put through one of these processes it is only a fraction of its original size. These much smaller files can then be sent quickly and easily over the Internet.

Files that have been packed together like this in **archives** are normally recognised by their file extensions, such as *.zip, .arj* or *.lha*. They have been shrunk using special programs (eg PKZIP or WinZip), and a corresponding program is needed to restore the compressed file to its original size. Files with the extension *.zip* can be unpacked easily using the program *WinZip*.

If you have not got one of these programs you can either get it from the cover CD-ROM of a computer magazine or from an FTP server. On FTP servers, look out for folders or directories that have the title 'Tools' or 'Utilities' – you will always find a decompression program in there. Or you can download WinZip from *http://www.winzip.com.*

After you have unpacked the file into its individual components, you can confidently delete the original compressed file.

Installing the FTP program under Windows 95

In the last step you 'broke down' the compressed file, also known as an archive, into its component files. Now you can start the installation procedure and install the FTP program on your computer. Click on the RUN item in the Windows 95 start menu.

In the window that appears now, you have to enter the name of the file concerned. If you have forgotten it you can browse through your hard disk and find the file that you loaded from the Internet.

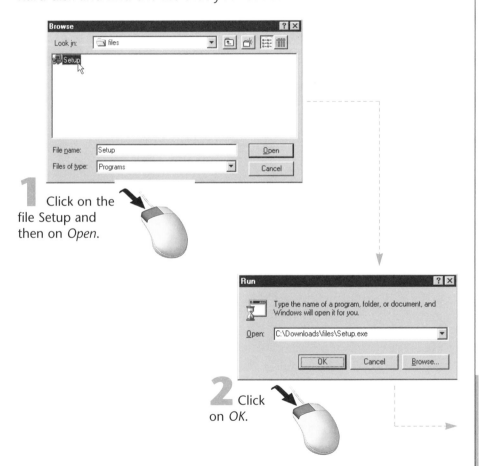

1 Click on the file Setup and then on *Open*.

2 Click on *OK*.

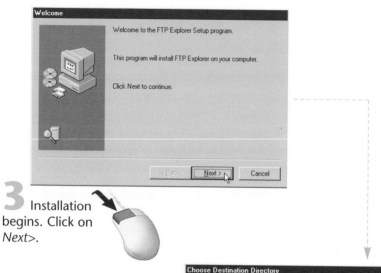

3 Installation begins. Click on *Next>*.

4 Enter a folder for the software to be installed in, or accept the folder that is suggested. Then click on *Next>*.

The following window requires you to make a few decisions. The first check box, *Create Shortcut on Desktop,* means that an icon will be shown on the Windows 95 desktop if the box has a tick in it. This icon, or **short cut**, saves you searching the Windows 95 menu structure for the program and saves a lot of time. The other check box, *Create Shortcut in Start Menu/Programs,* allows you to integrate a shortcut into the Windows 95 start menu.

The bottom check box will display a file with more information if you activate it. The default setting is that all the check boxes are activated. Leave them as they are.

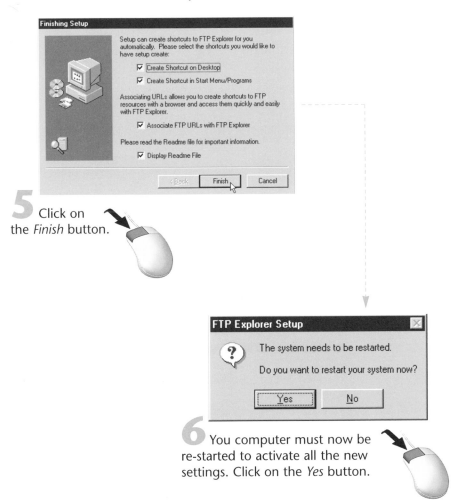

5 Click on
the *Finish* button.

6 You computer must now be
re-started to activate all the new
settings. Click on the *Yes* button.

After your computer and Windows 95 have been restarted you will
find a new icon on the Windows 95 desktop.

207

7 Click on the FTP
Explorer icon on the
Windows 95 desktop.

Using FTP to load a file
onto the computer

Well, now that you have everything you need, you can leap into
action with FTP. Your target is the Microsoft FTP server and you are
going to load a graphic from it onto your computer.

Click on the FTP program icon on your Windows 95 desktop.

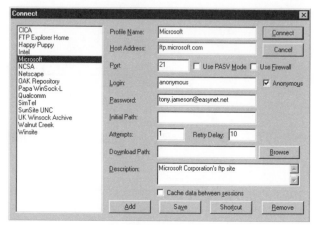

1 Click on Microsoft
in the left-hand field of
the window and then
on the *Connect* button.

After clicking on this button you are connected with the Internet (just like in previous chapters). As you will have noticed, our FTP program already contains the standard addresses of some FTP servers, so at this point you do not have to enter an FTP address.

When you use FTP to obtain a program, you are only loading its files onto your computer. The program is not ready to run straight away. You still need to unpack it and install it. In most cases there will be a file on your hard disk with the name 'Setup' or 'Install' which you must click on after downloading in order to install the program.

As soon as you are on the Internet, the FTP program connects to the Microsoft FTP server. You are connected to the **root directory** of the FTP server. From this position you can dig deeper into the structure of the FTP server until you finally reach your target.

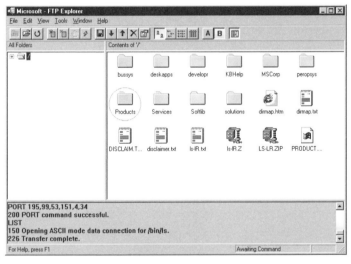

2 Double click on the index card entitled *Products* to open this folder.

The file that you want to download onto your computer is located in the folder *Products/Windows/Windows95/CDRomExtras/Funstuff*. To reach this folder you must keep on clicking down the hierarchy until you reach your target.

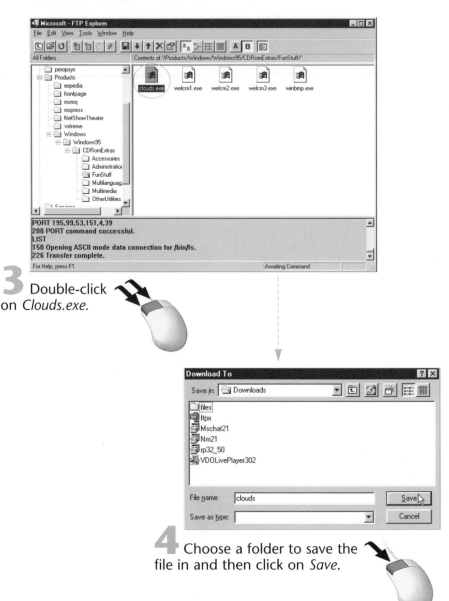

3 Double-click on *Clouds.exe*.

4 Choose a folder to save the file in and then click on *Save*.

When you have completed the last step, your FTP program starts to load the file from the FTP server. Once this is finished (it only takes a few minutes) you can leave the FTP program and close the Internet connection.

5 Click on the EXIT command in the FILE menu and leave the program.

If you now look in the folder that you gave as destination folder you will find the file *Clouds.exe* once again. You can unpack this program by running it through Windows Explorer or in a DOS window.

6 Voilà: the picture that you loaded from the Microsoft FTP server.

211

Entering a new FTP address

The next thing to learn is how to enter an **FTP address** within your FTP program. To do this, start up the FTP program and create a connection with the Internet.

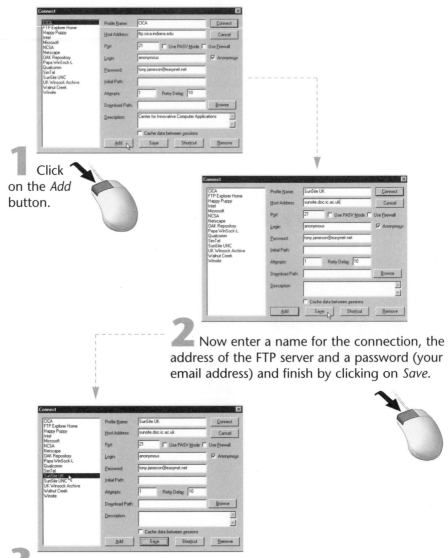

1 Click on the *Add* button.

2 Now enter a name for the connection, the address of the FTP server and a password (your email address) and finish by clicking on *Save*.

3 The new item: *Sunsite UK*.

Now, when you click on Connect you will connected to this huge and useful FTP server.

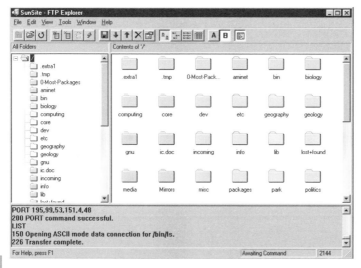

4 The connection with the new FTP server and its contents.

10

Discussions on the Internet: Newsgroups

What's in this chapter?

Newsgroups are one of the most popular Internet services, together with email and FTP. People from a wide variety of races, cultures and countries meet here and chat about all sorts of subjects. In this chapter you will learn where you can get a newsgroup program and how to install it. This will also give you an opportunity to learn a little about 'Netiquette'.

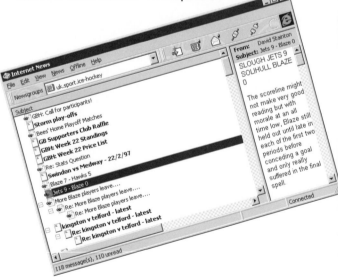

215

From A to Z – Newsgroups have something for everyone

As we all know, humankind is a very communicative species. The Internet with its newsgroups provides for this characteristic of ours. 'Usenet' is the largest of the electronic **discussion forums**. This giant source of information enables news to be exchanged within the whole of the Internet. In practice, it means that people from all over the world can take part in discussions on all conceivable subjects. The communities within **Usene**t that deal with particular subjects are called 'Newsgroups'.

Individual newsgroups can deal with all possible subjects from cars, cinema, sexuality and computers to the newsgroups themselves. At present there are approximately 15,000 newsgroups – all dealing with different subjects. You can see how very popular this method of communication is. Newsgroups provide a meeting place for anyone who has something to say on a particular subject, and would like the rest of the world to know about it too.

WHAT'S THIS?

A **Newsgroup reader** is a program which enables you to get connected to newsgroups and read, and reply to, the messages in them.

To take an active part in the life of a newsgroup you need a suitable program, known as a **newsgroup reader**.

There are moderated and unmoderated newsgroups on the Internet. In a moderated newsgroup all the messages that are sent to the newsgroup are first processed by a moderator. This person decides whether or not to publish the message in that newsgroup. In an unmoderated newsgroup all messages are automatically passed on to the whole of the newsgroup.

Messages that are waiting to be published in a newsgroup are sent from **Usenet servers** to all the locations that are allowed access to the newsgroups concerned. Normally only the most recent messages are displayed because the storage capacity of the newsgroups' servers is limited. As a rule, older contributions are archived and then later deleted. To have the very latest messages in a newsgroup displayed for you, you can have yourself put on a list, equivalent to having a subscription. The next time you visit that newsgroup the latest news is automatically displayed on your screen.

Multimedia data includes data such as sound, graphics and video sequences.

In addition to text, the messages in a newsgroup can contain pictures, sounds and other **multimedia data**.

How Usenet works

Usenet can be thought of as a globally available blackboard. Within this definition it acts as a discussion forum. To take an active part in what is going on, Internet users send and receive messages that are published in a newsgroup.

Newsgroups and the messages attached to them are administered on newsgroup servers distributed all over the globe. These servers are organised into **major divisions** which are themselves subdivided. All the newsgroup servers and the Usenet servers can communicate with each other, with the result that all messages that are published on one server are also available to all the other servers. As you gain experience with newsgroups, you will observe that not all servers publish all the newsgroups. Each newsgroup server decides which newsgroups it will look after and which ones it will not.

A wide variety of data such as pictures, sound or multimedia files can be published in newsgroups. This data must be specially **encoded** before it is published, so that it can be used. To view or play these files they first have to be transferred onto your computer and then decoded using special software.

The **newsgroup reader** enables you to read news and reply to it. Software of this kind also enables you to be put on newsgroup lists so that you always receive the latest messages. It is also possible to get yourself deleted from these lists.

There are thousands of newsgroups, classified hierarchically under special subject areas. The advantage of this is that it is easier to find newsgroups that interest you.

At the top of the newsgroup hierarchy is the major division. The following major divisions are currently available:

Description	Abbreviation
computer	COMP.
alternative	ALT.
social	SOC.
science	SCI.
recreation	REC.
news	NEWS.

The major divisions are divided into subsets such as *SCI.ASTRO*, which can be followed if necessary by a further subdivision: *SCI.ASTRO.HUBBLE.* This newsgroup, for instance, covers the Hubble telescope.

In most cases you can tell from the name of the newsgroup what subjects it deals with.

Downloading a newsgroup reader

Once you are familiar with the necessary theory, you are ready to install a newsgroup reader on your computer. Everything you need at this stage is available on the Internet. As before, you begin by visiting the Microsoft Web server and this has the relevant newsgroup reader. In this case the program offered by Microsoft is also able to receive emails (more about this later).

Connect yourself up to the Internet and start Internet Explorer. Enter the URL of the Microsoft home page: *http://www.microsoft.com/ msdownload.*

1 Click on the hyperlink *Internet Explorer Add-Ons.*

219

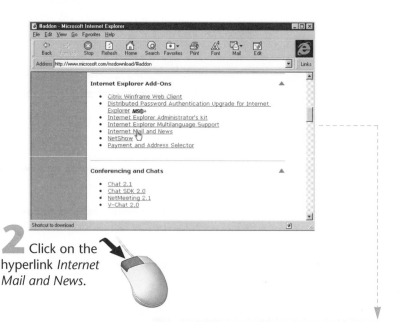

2 Click on the hyperlink *Internet Mail and News*.

3 Again, click on the *Internet Mail and News* link.

In the next few steps you have to specify the operating system under which the news program will run, in our case Windows 95, and the language.

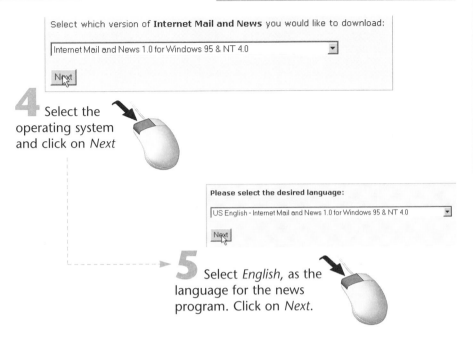

Select which version of **Internet Mail and News** you would like to download:

Internet Mail and News 1.0 for Windows 95 & NT 4.0

Next

4 Select the operating system and click on *Next*

Please select the desired language:

US English - Internet Mail and News 1.0 for Windows 95 & NT 4.0

Next

5 Select *English*, as the language for the news program. Click on *Next*.

In the next step you must establish an **Internet connection** to a Web server that has the file that you need to unload. As usual, choose the site closest to you, ideally a site in the UK if one is listed.

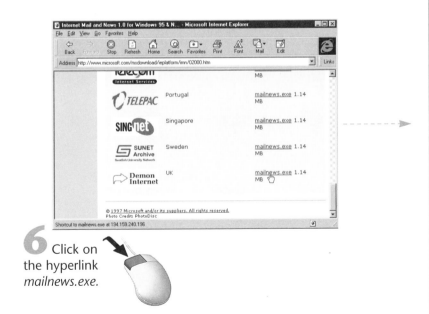

6 Click on the hyperlink *mailnews.exe*.

221

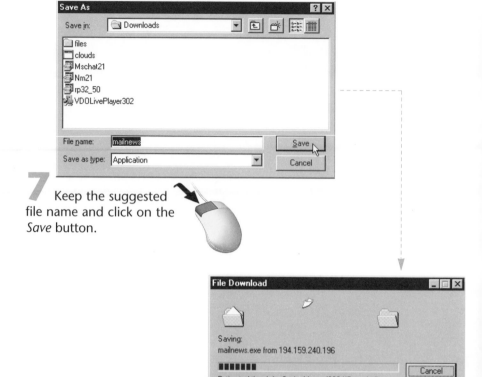

7 Keep the suggested file name and click on the *Save* button.

8 Wait a few minutes until the download is completed.

After a minute or two the news program is in the destination folder that you defined before the download. Now you can unpack the program and install it on your computer.

Installing the Microsoft news reader

Now that the compressed file is on your computer, the next thing to do is **unpack** it and install it. Open the start menu and select the item *Run*. Now enter the file name or look for the file on your **hard disk**.

1 Click on the file with the name *mailnews*, and then click on the *Open* button.

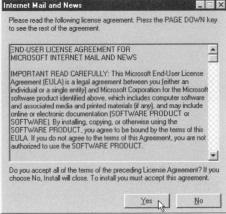

2 Once again it is time to accept the licensing conditions. Click on *Yes*.

223

3 All the files are now unpacked.
It is time to look at them.

4 Enter your
name and a
company name if
relevant. Then
click on *Next>*.

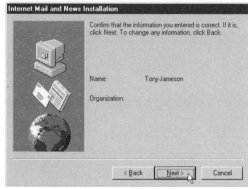

5 Click
on *Next>*.

6 Click on *Internet News* and then on *Next>*

7 Click on *Finish*.

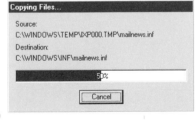

8 Wait until all the files have been copied.

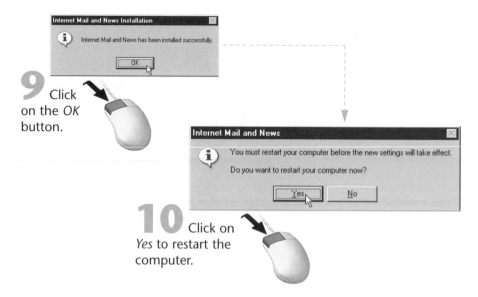

9 Click on the *OK* button.

10 Click on *Yes* to restart the computer.

In the last step of installation you have to restart the computer. Once the computer has started up again there is an item called INTERNET NEWS in the PROGRAMS menu of Windows 95. Click on this. You will then be required to configure the program.

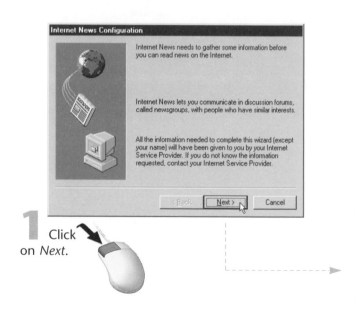

1 Click on *Next*.

2 Enter your name
and email address.

Now it is time to have a closer look at the documents you have
received from your Internet Service Provider. These should give you
all the information that you need at this stage. If anything is not
clear, a telephone call to your ISP should solve the problem.

At the next stage of installation, you need the name of the
newsgroup server. In this case it is *news.easynet.co.uk.*

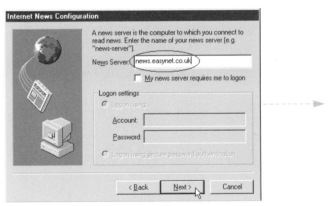

3 In the input field entitled *News Server*
enter the name of the news server concerned.

In the next step you will enter the type of **connection**. As a general rule you will be gaining access to the Newsgroups via a modem connected to the computer. To do this, you define a new connection to the **Internet Service Provider** which creates a connection with the Internet. In addition, it is possible to have access to an existing profile. However, you are going to define a new Internet connection. Why? Because practice makes perfect.

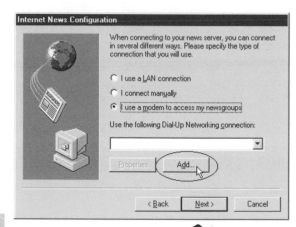

Click on the option field *I use a modem to access my newsgroups* and then on the *Add* button.

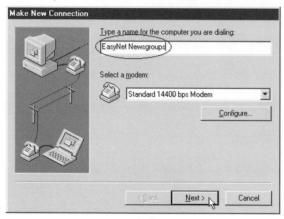

Enter a new name for the connection and select a modem. Then click *Next>*

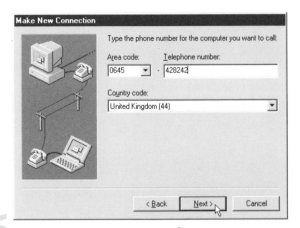

6 Enter the dialling code and telephone number and also the national code of your Internet Service Provider. Click on *Next>*.

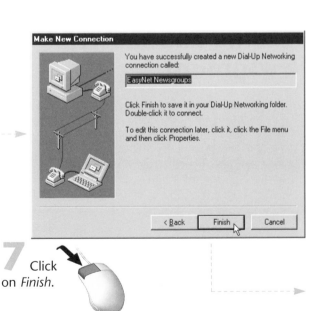

7 Click on *Finish*.

229

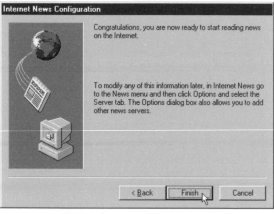

Internet News Configuration

Congratulations, you are now ready to start reading news on the Internet.

To modify any of this information later, in Internet News go to the News menu and then click Options and select the Server tab. The Options dialog box also allows you to add other news servers.

< Back   Finish   Cancel

8 Click on the *Finish* button.

Reading and subscribing to Newsgroups

We have now reached the point where you can take an active part in the life of the Newsgroup. So, start up the Newsgroup Reader. At this point the program notes that you have not yet subscribed to any Newsgroups. The Newsgroup Reader asks you whether you would like to do so. The program will connect up with the Internet and load a list of available **Newsgroups** on your computer. To proceed, try the Newsgroup server of the appropriate Internet Service Provider – in this case Easynet.

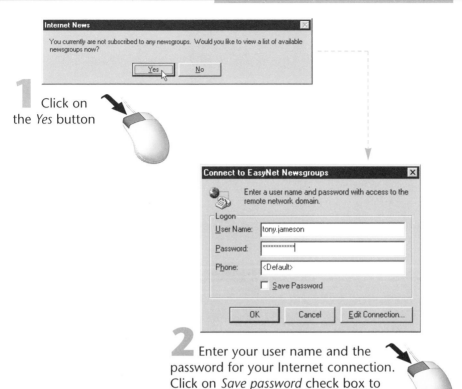

1 Click on the *Yes* button

2 Enter your user name and the password for your Internet connection. Click on *Save password* check box to save the password. Close all the inputs by clicking on the *OK* button.

As soon as you click on the *OK* button your computer connects to the Internet. After the *connection* has been successfully made, all the Newsgroups that are available on that Newsgroup server are downloaded. This takes a little time. Depending on the Internet Service Provider, 10,000 to 15,000 Newsgroups may be loaded.

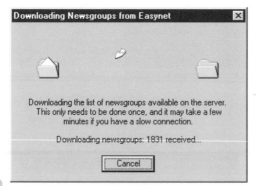

3 The list of available Newsgroups is loaded. You must exercise a bit of patience at this stage.

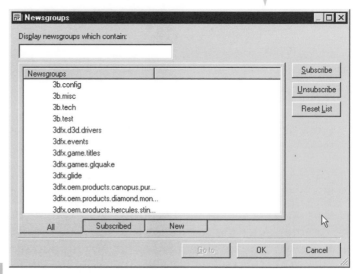

4 All the Newsgroups are now available for you.

Now you have a list of all the Newsgroups on your computer. If you wish to look for a particular Newsgroup you can use Search facilities.

When you have discovered a Newsgroup that looks interesting, you can subscribe to it. All you need do is double-click on the Newsgroup concerned.

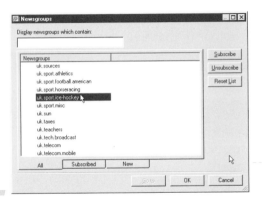

5 To subscribe to a Newsgroup that you like the look of, double-click on it. Then click on *OK*.

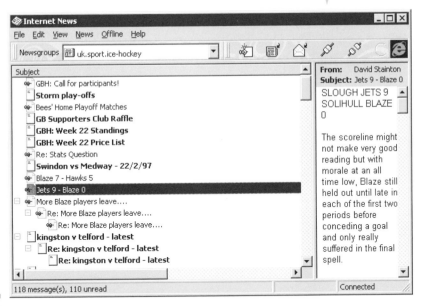

6 Click on an entry to look at its contents.

If you need to respond to an item, call it up and click on the *Reply to Author* icon on the toolbar.

233

To post your own question within a Newsgroup, click on the icon marked *New Message* in the toolbar of the Newsgroup reader.

'Netiquette'

Obviously you are expected to be **polite** on the Internet. Just as in real life, you meet real people from all over the world on the Internet, not just bits and bytes. So here are a few tips on this particular form of communication.

It's good practice to 'lurk' for a while before sending your own messages to newsgroups. 'Lurking' is an Internet word for reading messages and getting the flavour of the group before contributing to it. And don't 'spam' (another Internet word, which means sending the same message to countless newsgroups regardless of whether it's relevant to that group).

The Internet is very friendly – you find that you are among like-minded people. You speak to each other as if you were in the family. The **rule of thumb** is, write as you would expect to be written to.

'Smileys' are often used in Internet messages. They are also called **emoticons**. They are combinations of letters and characters which are used to convey emotions within an email conversation, a chat or a Newsgroup message. In the Internet there are many lists of smileys. You can find a smiley for any occasion.

Internet surfers are masters of abbreviation. A lot of common expressions from everyday speech have been brought onto the Internet. To save writing these out in full every time, a list of abbreviations has been compiled. Some of these are included in the following table:

Shorthand	Meaning
(bg)	big grin
(g)	grin
ASAP	As soon as possible
BTW	By the way
FYI	For your information
FYEO	For your eyes only
HHOJ	Ha ha only joking
IDU	I don't understand
L8R	Later
LOL	Laughing out loud
PITA	Pain in the a***
ROFL	Rolling on floor laughing

There are lots more of these abbreviations. Have a rummage around in the Internet. It won't take long to find some.

11

Sending
an email

What's in
this chapter?

This chapter will introduce you to a classic
Internet service – email. You will learn about
how it works and the different types of
email. Then you will get an email program
from the Microsoft
Web server that is
installed under
Windows 95.

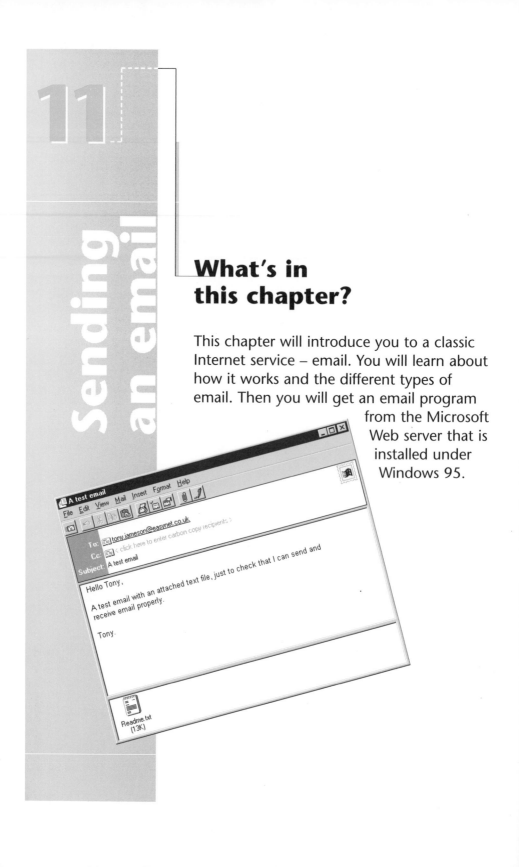

You already know:

You are going to learn:

What is an email?

The postal service is a method of communication that has existed since ancient times. People happily use it to converse with each other. But today, the electronic post, known for short as email, has largely overtaken the traditional method of sending letters. There are many reasons for this. First of all, it usually takes only a few minutes for an email to get to the other end of the world. Moreover, it is cheap and has a single uniform **rate**. People in many companies and even in their private lives are using the electronic post more and more. Every day, millions of emails whizz round the globe. Email is an outstanding tool for communicating with relatives, acquaintances and friends and colleagues from the world of work, and keeping in touch with them.

CAUTION

Emails are an ideal way of transmitting viruses. If ever you receive an email with a chaotic set of characters in either the subject line or the address line, do not open it. It could contain a virus that is automatically activated as soon as you open the email.

Unlike the normal postal service (sometimes referred to as 'snail mail'), email allows you to 'attach' graphics, sounds and **video messages** to your electronic letter.

Sending an email works like this: the message is first broken down into small individual data packets. Each of these convenient little packets is given a 'sticker' which contains the recipient's address. In the Internet there are **routers** which send the packets to the recipient by the fastest route.

WHAT'S THIS?

An **Internet router** is a computer in the Internet that forwards data from you to the Internet and to you from the Internet. Every Internet Service Provider has Internet routers.

Individual packets can even be sent by different routes if one route through the Internet is suddenly no longer available. If this happens, the Internet router looks for an alternative path and uses that to send the remaining **data packets**.

When all the packets arrive at their destination they are reassembled to form the complete email. This is not exactly how it happens in practice. The emails actually arrive at the Internet Service Provider which collects all the email for a particular person. When you create a connection to the Internet, all your personal emails are loaded onto your computer in a single operation.

Within the Internet, emails can also be exchanged between all the large online services. This means that you can send emails from the Internet to a subscriber in an online service such as AOL, CompuServe, or T-Online.

email addresses

To send an email you need an email address. These addresses vary depending on the Internet Service Provider. For instance, my email address is *tonyjameson@easynet.co.uk*. In front of the 'funny' sign, which is also known as the **'at'** sign, is my personal name. After it, comes the address, also known as the domain. Your Internet Service Provider will supply you with an email address. As a general rule you are not entitled to choose your own.

In online services the whole subject of addresses has a slightly different aspect. In CompuServe, for example, every user is given a user ID. This is composed of figures separated by a comma, such as 123456,789. Within **CompuServe** email always comes and goes using this number. However if, from the Internet, you want to send an email to a CompuServe user, you must put a different address. This is as follows: 123456.789@compuserve.com. Very important, the user ID must be separated by a dot instead of a comma.

In **AOL** you have the option of choosing your own email address, provided that the name is not already in use. For instance, my AOL name is *tonyjames*. As in CompuServe, it is sufficient to use just this name for communication within AOL. However, if you wish to send an email to my AOL address from the Internet, the syntax must be changed. Just as for CompuServe, you can work the address out for yourself *tonyjames@AOL.com*.

With T-Online the user name consists of the telephone number plus the code. After this comes the **user identification**, which is separated by a dot. If you wish to send an email to T-Online from the Internet you must enter the following syntax: 022112345.001@t-online.com.

How to obtain an email program

Windows 95 already has an integral mail program – Microsoft Exchange Client. However, you are going to fetch a different (and much better) program from the Internet. This is on the Microsoft WWW server and is perfect for our purposes.

Make an Internet connection and, in the web browser, enter the following URL: *http://www.microsoft.com/msdownload.*

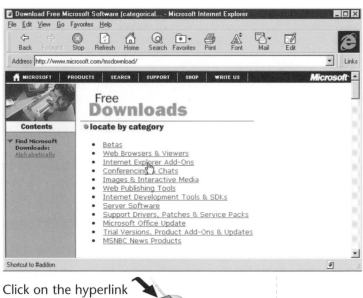

1 Click on the hyperlink
Internet Explorer Add-Ons.

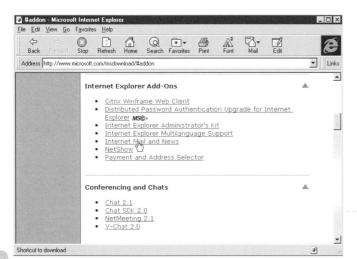

2 Click on the hyperlink *Internet Mail and News*.

3 Click on *Internet Main & News* again.

241

Select which version of **Internet Mail and News** you would like to download:

| Internet Mail and News 1.0 for Windows 95 & NT 4.0 | ▼ |

Next

4 Select your operating system (*Windows 95*), and click on the *Next* button.

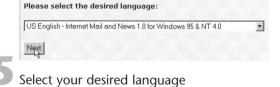

Please select the desired language:

| US English - Internet Mail and News 1.0 for Windows 95 & NT 4.0 | ▼ |

Next

5 Select your desired language (English), and click on *Next*.

In the next step of installation you must select the connection that you wish to use to load the file from the Internet. Choose the server located geographically closest to you as this usually provides the quickest download time.

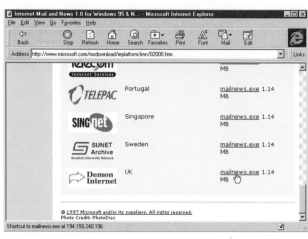

6 Click on one of the available connections to load the file *mailnews.exe*.

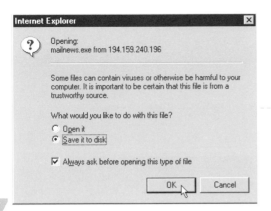

7 Click on the *Save it to disk* button and then on the *OK* button.

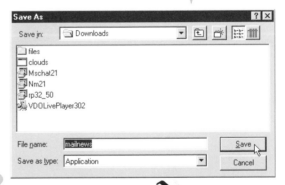

8 Select a folder into which the file is to be copied. Then click on the *Save* button.

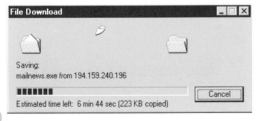

9 Wait a little while until the file has been loaded onto your computer.

After a few minutes, the file should be on your computer and available for your use.

243

Installing the Microsoft email program

Once the compressed email file is on your hard disk you will learn how to install the program so that you can use it.

Open the Windows 95 Start menu and select the menu item RUN.

Enter the path to the downloaded file or click on the *Browse* button to find the file. Clicking on *OK* then starts the installation.

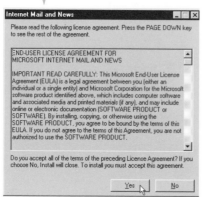

Click on *Yes* to accept the terms of the licence agreement.

The file that has been downloaded from the Internet is unpacked. Wait a few moments.

4 Enter your name and, if applicable, the company name. Then click on *Next>*.

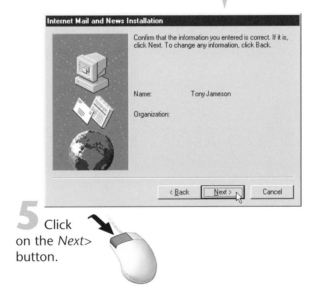

5 Click on the *Next>* button.

In the next step of installation the program will need to know which components you would like to install. The program that you have loaded from the Internet is a combined email and news-reader program. At this point you will only install the email program.

245

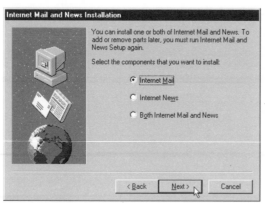

Internet Mail and News Installation

You can install one or both of Internet Mail and News. To add or remove parts later, you must run Internet Mail and News Setup again.

Select the components that you want to install:

- ⦿ Internet Mail
- ○ Internet News
- ○ Both Internet Mail and News

< Back   Next >   Cancel

6 Click on the *Internet Mail* radio button and then on the *Next>* button.

Internet Mail and News Installation

Setup will now copy the necessary files onto your computer and update settings. This may take a couple of minutes.

After all of the files have been copied, you may need to restart your computer before you can use Mail or News. If a restart is required, Setup will prompt you.

< Back   Finish   Cancel

7 Click on *Finish*.

Copying Files...

Source:
C:\WINDOWS\TEMP\IXP000.TMP\mailnews.inf
Destination:
C:\WINDOWS\INF\mailnews.inf

50%

Cancel

8 The files are now being installed on your computer. Wait while is done.

9 Click on
OK button.

After you have been informed that the program has been
successfully installed you must restart your computer. This will
activate all the newly installed files.

10 Click on
the *Yes* button.

Sending an email

Before you can go ahead and send an email you must configure with
your personal Internet
settings. When you have
started the program for the
first time an 'installation
wizard' accompanies you
and helps you with the
configuration.

> Do not reveal in an email any personal
> information that is not for public
> consumption. Internet users have plenty
> of opportunity to intercept your emails
> and read them. Encryption programs,
> which encode your email, can be helpful
> in this respect.

Open the Windows 95 Start menu and select the PROGRAMS item. In the submenu you will see an item called INTERNET MAIL. This opens our newly installed email program.

1 Click on *Next>*.

In the following window you must enter your name in the first field. This will then be used to inform the recipient who has sent their incoming email. In the second field the email program requires you to enter your email address. If you do not know this, the best thing to do is to call your Internet Service Provider. They will certainly be able to help you.

2 Enter your name
and your email address.
Then click on *Next>*.

In the next step you must input the **address** of the Internet
computer that delivers your email. You can get this information
either from the accompanying literature on **Internet access**, or you
can call your Internet Service Provider for further information.

In the case of the Internet Service Provider Easynet, the relevant
computer is *mail@easynet.co.uk*.

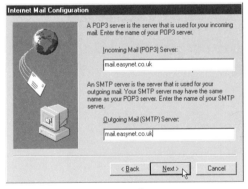

3 Enter the name of the
mail computer. Click on *Next>*.

249

You will need to consult the documents supplied by your **Internet Service Provider** again for the next stage of installation. You must now enter the user name and **password** which give you the right to collect your emails from your Internet Service Provider.

Enter the user name and your password and click on *Next>*.

Now you have to define the type of Internet connection. After this, you activate an existing data communications connection in the Internet.

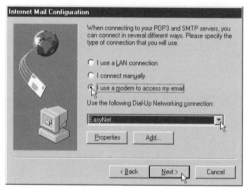

Click on *I use a modem to access my email*. In the list box choose an Internet connection that has already been created. Finish this step by clicking the *Next>* button.

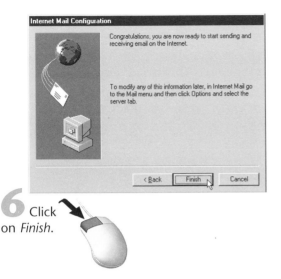

6 Click on *Finish*.

Done! The email program is now installed on your computer. Open the Windows 95 Start menu and in the PROGRAMS menu select the item INTERNET MAIL. The email program opens.

To create a connection with the Internet Service Provider, click on the *Send and Receive* button. When this is done the email program contacts the Internet Service Provider and collects any email that is waiting for you.

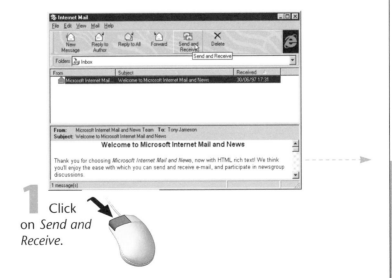

1 Click on *Send and Receive*.

251

2 Enter your user name and password. Then click on *OK*.

An important point is still missing – sending an email. Start by sending an email to yourself. This way, you will both send and receive a message, so you will know that everything is working properly!

Begin by starting up Microsoft Internet Mail with the aim of sending an email.

You are also going to attach to your email a text file that you have written in Notepad, which will also be sent with the email.

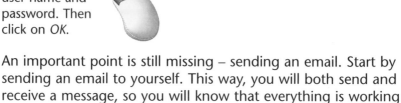

1 Click on INTERNET MAIL in the Start menu.

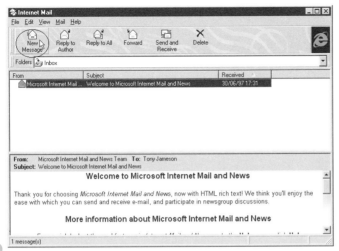

2 Click on the *New Message*
button in the toolbar.

When you click on the **New Message** button, a new window
appears where you can create a message.

In the *To:* field you must enter the email address of the **recipient**
(in this case your own email address). The *Subject:* field normally
contains a brief description of contents of the email. This will appear
as a summary in the recipient's email program Inbox.

To enclose a document with an email you click on the **paperclip**
icon. Then you can select a document from your hard disk.

3 Enter the email address of the recipient, the subject line and the text of the email. Add to this a document that is to be sent with your email. Click on the envelope icon in the toolbar.

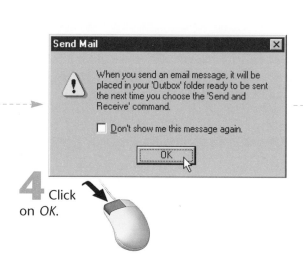

4 Click on *OK*.

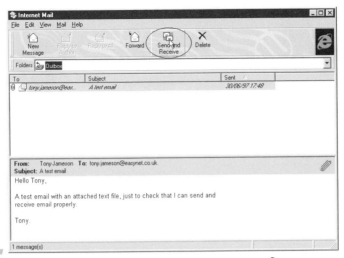

5 Your email is now in the outgoing post box.
Click on the *Send and Receive* icon on the toolbar.

6 Click on the *OK*
button to connect to
the Internet.

As soon as the connection is made, Internet Mail will send this
message (along with any others waiting in the Outbox) and collect
any messages addressed to you. Your incoming messages will appear
in the Inbox. The message you sent to yourself in the steps above
should be received immediately.

255

12

Your own home page on the WWW

What's in this chapter?

Now that you're familiar with the details of the Internet services you can really move into the Internet way of life. You are going to set up a simple Web page and in doing so learn a few HTML commands. Once you have done this, there are a number of ways of publishing your page on the Internet. You will find out how you can publish your Web page in an online service such as CompuServe or AOL.

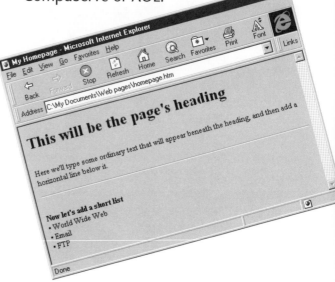

You already know:

You are going to learn:

You only need to know a few basics

These days it is almost 'the done thing' to have one's own home page on the Internet. Most online services and a large number of Internet Service Providers offer hard-disk space for hire on their Internet computers, where you can put your own home page. You may think this would only be for absolute computer nerds and professionals. You are wrong. You do not need a lot of knowledge to create your own **home page**. In some cases you don't even need to pay to do it. More and more ISPs now give you this hard-disk space for free!

Avoid putting large graphics on your home page. The larger these are the longer a person visiting your Web site needs to wait while these load and he or she may very soon get bored and quit.

You do not need any special knowledge to create a home page. However, it is a good idea to surf a few times on the Internet to get the feel of various WWW pages.

The language that is used to create Web pages is called HTML. Written out in full this is **Hypertext Markup Language**. It is an Internet program language which runs on all computer platforms and gives roughly the same results on any Web browser screen.

Very likely you have little idea of HTML at the moment, but that will soon change. There are dozens of programs on the Internet to help you create HTML pages. However, the text editor in Windows 95 will be sufficient for our first attempt.

Creating your own home page

Call up the Windows 95 *editor* as follows:

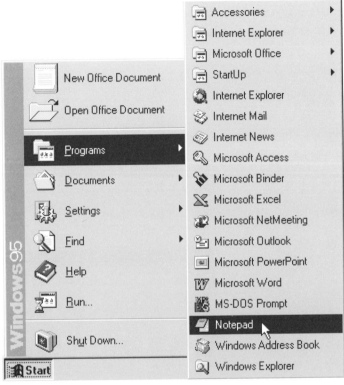

1 In the start menu, select the item NOTEPAD.

You will now type a short HTML program into the text editor. The commands you need to put at the beginnings of the lines have the following meanings:

<H1>	produces a heading
 	starts a new line
<HR>	produces a horizontal line
<LI>	produces a list
<B>	switches on bold face

259

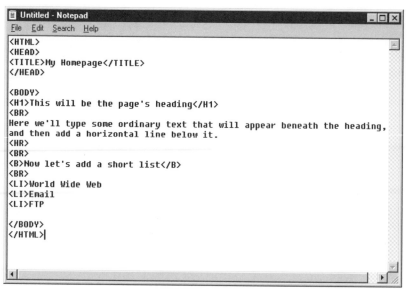

2 Type in the whole of the little program shown here.

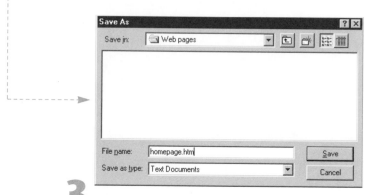

3 Save the program that you have entered under the name *homepage.htm*

Now start up Microsoft Internet Explorer, but without connecting to the Internet. Instead, call up the Web page that you have just created within Internet explorer.

As soon as Internet Explorer is called up, open the FILE menu and click on the OPEN menu item.

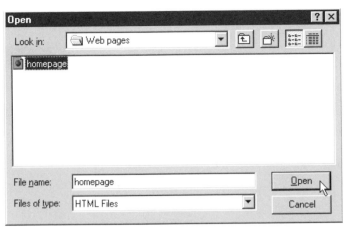

4 Open the new Web page by selecting the file and clicking on the *Open* button.

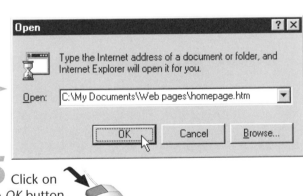

5 Click on the *OK* button.

261

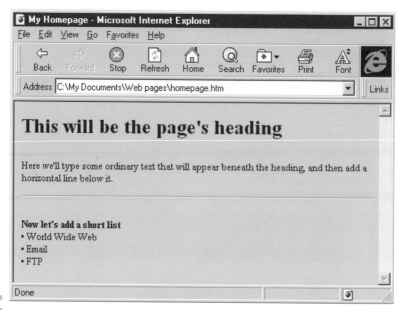

This will be the page's heading

Here we'll type some ordinary text that will appear beneath the heading, and then add a
horizontal line below it.

Now let's add a short list
- World Wide Web
- Email
- FTP

 Voilà! your very own home page.

A home page in an online service

Well, we now have a **home page**, but this is not on the Internet,
only on your hard disk. The online services AOL and CompuServe
offer their subscribers the facility to publish Web pages as part of
their service. Of course, in the last step we only scratched the
surface. If you want to go deeper into the subject, there are excellent
workshops on the Internet which pass on basic knowledge. For
example, you can learn how sound and video sequences can be
integrated into a Web page.

In **AOL** you have the opportunity to put up to 2Mb of data onto the
Internet if you are a member. The people at AOL who are responsible
for this expect you to send your complete Web page, including all
graphics, by email. When they have received all the data at AOL it
only takes a few days for your Web page to be accessible via the
Internet. The syntax is then as follows:
http://www.aol.home.com/_AOL_Name/index.html.

The procedure for **CompuServe** follows the same pattern. Just telephone the online services and get them to send you the documentation.

In the meantime there are a lot of Internet Service Providers which also offer private individuals the opportunity to make their information available on the Internet. The prices vary depending on the scope of the Web pages and the data traffic situation.

It's just like real life: only a few things work properly the first time and even then they don't always do so.

For this reason, I have picked out a few of the typical problems that you might encounter during the course of your Internet career by way of example.

The modem will not dial

In most cases the reasons for this are very clear. Mostly, it is simply forgetting to switch the modem on, or because the connecting cable between the computer and the modem is not properly pushed in, is defective or perhaps is just not there. If possible, always use the manufacturer's original cable.

If you still have no success, check that there is electricity is flowing through the wall socket.

Another common error is that you have the wrong modem installed in Windows 95. Windows 95 only works with modems that it recognises and that are installed.

You cannot get onto the Internet

If the modem is working and is dialling, but you still cannot get connected to the Internet, check the telephone line. If there is another telephone connected to the same line as your modem, it is possible that somebody else is using it.

Check that you have set the correct dialling procedure. There are two possible settings: pulse dialling and tone dialling. With pulse dialling, you can hear 'clicking' as you dial, with tone dialling you hear electronic notes.

If your telephone is connected to a switchboard, you must dial a number to get a line before you can dial the Internet number. This number varies from switchboard to switchboard.

If all else fails, then check the telephone number. Perhaps a couple of digits have been transposed and the wrong number is being called.

An existing Internet connection is cut off

This is particularly irritating, especially if you are just about to finish a long download. In most cases the reason is that the interference on a line is too intense. There is only one remedy for this – dial again and hope that the new connection is better.

A URL is not accepted

There are many reasons why a Web browser may be unable to find a particular URL. Check, first of all, that a typing error has not crept in. In very many cases it is due to a network fault. Perhaps the WWW server concerned is having maintenance work done. It is also possible that the WWW server is so busy that it is completely overloaded.

A Web page cannot be found

One of the commonest messages you will come across during your Internet career is the one with the cryptic wording '404 not found'. The explanation for this is simply that this page no longer exists. This can happen if WWW pages are altered because they have become out of date, or have been renamed or deleted altogether. Just accept it.

America Online America Online was launched in 1985 and currently has several million users throughout the world. It is currently the largest online service in the world.

Anonymous FTP Logging on to an FTP server by entering your name as the user name and your email address as the password.

AOL The European online service which is maintained by America Online and Bertelsmann and is modelled on the American service of the same name. However, a lot of its content is local in character, serving for example, the UK, France and Germany.

Baud or baud rate. Term used for the signalling rate of a modem, that is the number of signal changes occurring each second that a modem can process. The higher the baud rate, the faster the modem.

Chat The facility that lets you communicate with other users online. Subscribers to AOL, for example, use Chat Rooms to discuss a wide variety of subjects. Live chats are also popular. They enable their members to put questions to celebrities in real time.

CompuServe An online service of American origin which also acts as an Internet Service Provider. To access this service you need CIM software. There are various versions, (DOSCIM for DOS, WinCIM for Windows, etc.).

DNS A process that assigns cryptic IP addresses to meaningful names and thus helps to call up a computer that is located within the Internet or a resource that can be reached through it.

DNS Server A computer owned by an Internet Service Provider which stores the IP addresses with their respective names. A lot of Internet service providers have a secondary DNS server as well as their main one, so that if the primary DNS server goes down or cannot be reached the service provider can still use DNS.

Download During a download an AOL user loads a file or a program from a particular AOL software library.

email Electronic messages that are sent within a local network, via online services or via the Internet. An email program or an email client that is connected to the relevant server is needed to process emails.

email address Address under which you can be reached in a particular email system.

Freeware Software that can be used free of charge. This facility often only extends to private use, with commercial users being liable to pay for it.

FTP Internet service for file transfer. An FTP client is needed for access. This downloads the files from an FTP server or makes them available on the server.

Gateway Converts data between different formats.

Home Page The primary Web page of a supplier on the World Wide Web.

Host General term for a computer, usually a large one, that is responsible for a particular task.

Hyperlink Link to another resource, usually on the Internet.

HyperText Document format in the WWW which applies not only to text but also to graphics, sound files and videos. It also includes hyperlinks.

Internet International Public Network that was first used in the USA for military and scientific purposes. Anyone can supply information for the Internet for, unlike an *online service*, there are no operators who own the Internet and its access points. Access to the Internet by modem or ISDN is provided by Internet service providers and online services such as AOL.

Internet Service Provider Company that supplies access to the Internet for private and commercial users and enables them to use its own channels for connecting to the Internet itself.

IP *See* TCP/IP.

IP address Unique address of a device within a network that runs under TCP/IP. Every computer on the Internet, for example, has a unique IP address. These addresses are issued by international organisations which ensure that every IP address is unique. A static IP address is always the same, while a dynamic IP address is allocated by the Internet service provider from a series of IP addresses made available for the purpose of dialling into the Internet.

ISDN Digital telephone network that integrates all speech and data services. The outstanding feature of ISDN is its high transfer rate of 64,000 bps and the rapid creation of a connection with the recipient, which takes place within one or two seconds.

Modem Artificial word made from *Modulator* and *Demo*dulator. Device that converts digital signals from the computer into analog signals, or sounds, for transmission over the analog telephone network, and vice versa.

Mosaic Ancestor of all WWW browsers. Also used as a synonym for any program of this type.

Netiquette Rules of behaviour (Internet etiquette) for electronic communication by email and in newsgroups.

Network Group of computers that are connected together. If the network is in one office or building it is called a LAN or local area network. Other types of network include online services (private networks) and WANs, or wide area networks, which link geographically separated local networks.

Newsgroups Part of the Internet in which all possible (and improbable) subjects are discussed. Communication takes place offline – users can leave messages for other users to read and answer. A newsgroup reader is needed to access newsgroups. On AOL

this is already available within the software package.

Online service Private network, national or international in character, to which customers connect via a modem or ISDN, for example. They can call up data that has been prepared by individual suppliers and made available for downloading. Access to an online service is via the points of presence provided by the operator concerned.

Password Secret identifier used in association with a user name. It protects access to the associated resources and is intended to recognise authorised users only.

PC card Bus system that works with credit-card sized expansion cards (formerly also known as PCMCIA cards).

PCMCIA *See* PC card.

Points of Presence Access points maintained by a mailbox and online service administrator or by the Internet Service Provider to which you can dial in using a modem or an ISDN line.

Pulse dialling Dialling method for analog telephone lines, in which the number being dialled is converted, digit by digit, into pulses of varying length. This method of dialling is considerably more time-consuming than tone dialling.

Shareware Try-before-you-buy software, which you can use free of charge for a certain length of time, after which, if you like it, you have to buy it.

Surfing Colloquial expression for the practice of visiting Web pages and Web servers, and clicking with the mouse on hyperlinks in order to jump to particular resources.

TCP/IP Network protocol used in network environments that encompass a number of systems, UNIX and above all in the Internet. Often also referred to as IP for short.

Telnet Internet service (terminal emulation)

Terminal emulation Simulates a terminal on the PC in order to establish contact with another computer (mainframe, mailbox, etc.) and be able to exchange inputs and outputs.

Tone dialling Dialling method for analog telephone lines, in which the number being dialled is converted, digit by digit, into notes of varying frequencies. This method of dialling is considerably less time-consuming than pulse dialling.

Upload Sending files to a mailbox, an online service or an FTP server.

URL Locates a resource on the Internet.

Usenet *See* Newsgroups.

WWW Graphical service on the Internet which permits access to a wide variety of resources (documents, files, videos, etc). A Web browser is needed to access this service.

Web browser A program with which the PC can access the Web servers available on the World Wide Web and also use other Internet services (FTP, Gopher, etc.) in addition to the WWW. A Web browser is provided as part of AOL version 2.5 software.

World Wide Web *See* WWW.

E

Easynet 71, 73, 230
email 30–4, 98, 236, 267
    address 144, 239–40, 267
    cost 31
    installing a program 244–7
    mailing lists 31–2
    multimedia messages 31
    obtaining a program 240–3
    password 250
    sending 98, 247–55
emoticons 234–5
encoding 218
enter key 13
escape key 14
etiquette, online 234
Explorer software see Internet
    Explorer

F

Favorites 98
fax-modem 52
fax software 52
file ending 100
File Transfer Protocol see FTP
file type 100
Find Next 103
Font 98
forums 35
Forward 97
Freeware 2, 37, 267
FTP 36–7, 198, 267
    downloading a program 201–3
    how FTP works 200–1
    installing an FTP program
    205–8
    loading a file onto the
    computer using 208–12

    root directory 209
    unpacking 204
FTP address 212–13
FTP Client 200
FTP Daemon 200
FTP server 37, 41, 202
function keys 14

G

gateway 268
Gopher 42, 120–35
    downloading 122–8
Gopher home server 42, 128
Gopher server 42, 121, 128, 131
Gopher Space 121, 128–31
graphics 39

H

Hayes-compatible modem 51
history 22–5
home page 27, 268
    creating 258, 259–61
    in online service 262–3
host 268
HPCA 25
HTML 104
HTML format 100
HTML page 104, 110
http 88
hyperlink 40, 98, 268
HyperText 40, 268
HyperText Markup Language see
    HTML
HyperText transfer protocol 88

L

LAN 22
licence conditions 80
link 40
Local Area Network (LAN) 22
login name 70
Lycos 111, 114–17

M

Mail 98
mail, electronic see email
mailbox 45
mailing lists 30, 32
    messages 33
    regulating 33
mainframe 22
match case 102, 103
microphone 184, 185, 193
Microsoft Chat 175
Microsoft home page 27, 97
Microsoft Internet Explorer 3.0 see
    Internet Explorer
Microsoft Network 44
MILNET 24
modem 48, 50–1, 68, 269
    AT standard 52
    configuration 55–61
    detecting 59
    driver 59
    fax 52
    Hayes-compatible 51–3
    installing 55–61
    location information 61
    secondary system 59
    setting up 55–61
    speed 50–1
    standard modem 59
    transmission rate 51
    type 60
    voice 53
    Windows 95 and 54–61
modem manufacturer 59
Modem Wizard 54, 56
moderated discussion groups 33
modulator 50
Mosaic 269
mouse 17
MSN 44
multimedia data 217

N

NASA 24
National Research and Education
    Network (NREN) 24, 25
National Science Foundation (NSF)
    24
navigational keys 15
Navigator 39
Netiquette 234, 269
netMeeting 184
    directory of users 195
    installing 185–94
Netscape 78
Netscape Navigator 39, 146
network 22, 68, 269
network cable 23
network card 64
network configuration 63
network protocol 63
new settings 72
News 151
newsgroup reader 34, 216, 218
    downloading 219–22
    installing 223–30
newsgroups 34–5, 98, 214, 269
    access to contents 98
    article 34